Influence of flight control laws on structural sizing of commercial aircraft

TU Braunschweig – Niedersächsisches
Forschungszentrum für Luftfahrt

Berichte aus der Luft- und Raumfahrttechnik

Forschungsbericht 2021-19

Influence of flight control laws on structural sizing of commercial aircraft

Rahmetalla Nazzeri

TU Braunschweig
Institut für Flugzeugbau und Leichtbau

Diese Arbeit erscheint gleichzeitig als von der Fakultät für Maschinenbau der Technischen Universität Carolo-Wilhelmina zu Braunschweig zur Erlangung des akademischen Grades eines Doktor-Ingenieurs genehmigte Dissertation.

Bibliografische Information der Deutschen Nationalbibliothek
Die Deutsche Nationalbibliothek verzeichnet diese Publikation in der Deutschen Nationalbibliografie; detaillierte bibliographische Daten sind im Internet über http://dnb.d-nb.de abrufbar.
1. Aufl. - Göttingen: Cuvillier, 2021
Zugl.: (TU) Braunschweig, Univ., Diss., 2021

Diese Arbeit erscheint gleichzeitig als von der Fakultät für Maschinenbau der Technischen Universität Carolo-Wilhelmina zu Braunschweig zur Erlangung des akademischen Grades eines Doktor-Ingenieurs genehmigte Dissertation.

Herausgeber der NFL Forschungsberichte:
TU Braunschweig – Niedersächsisches Forschungszentrum für Luftfahrt
Hermann-Blenk-Straße 42 • 38108 Braunschweig
Tel: 0531-391-9822 • Fax: 0531-391-9804
Mail: nfl@tu-braunschweig.de
Internet: www.tu-braunschweig.de/nfl

Nonnenstieg 8, 37075 Göttingen
Telefon: 0551-54724-0
Telefax: 0551-54724-21
www.cuvillier.de

1. Auflage, 2021
Gedruckt auf umweltfreundlichem, säurefreiem Papier aus nachhaltiger Forstwirtschaft.

ISBN 978-3-7369-7516-3
eISBN 978-3-7369-6516-4

Influence of flight control laws on structural sizing of commercial aircraft

Von der Fakultät für Maschinenbau der
Technischen Universität Carolo-Wilhelmina zu Braunschweig

zur Erlangung der Würde eines
Doktor-Ingenieurs
(Dr.-Ing.)

genehmigte

Dissertation

von:

M.Eng. Rahmetalla Nazzeri

aus (Geburtsort): Hamburg

eingereicht am: 04.03.2021
mündliche Prüfung am: 05.10.2021

Gutachter: Prof. Dr.-Ing. P. Horst
Prof. C. Armstrong, PhD

2021

Acknowledgement

The developments in the frame of this PhD thesis are the result of a collaboration between the Institute of Aircraft Design and Lightweight Structures of the Technische Universität Braunschweig and the Flight Physics department at Airbus in Hamburg.

In this sense I am very thankful to Prof. Dr.-Ing. Peter Horst and Dr.-Ing. Matthias Haupt who supported me during the time of this collaborative research project with their broad knowledge and experience.

Further I would like to thank Prof. Cecil Armstrong with whom I had several inspiring meetings during the time of this PhD thesis.

In addition, I am very thankful to Frank Lange, who was a personal mentor with his professional expertise in all matters of loads analysis, and Dr.-Ing. Michael Kordt, who managed the funding and working conditions with his valuable management skills.

Further I would like to thank Dr.-Ing. Stephan Adden for his wisdom and expertise during this PhD thesis.

I would like to express my special thanks to Christophe Sebastien who originated the project that laid the foundation for this PhD.

Finally I am very thankful to my family who supported me with patience during this period of my working career.

Abstract

The increasing demand for new civil aircraft across established airlines in industrialized countries and emerging markets pushes aircraft manufacturers like Airbus to develop innovative solutions that lead in particular to mass reductions, improved lift to drag ratios, higher take-off weight, less fuel consumption and hence less carbon dioxide emissions. One way to achieve these kinds of improvements more effectively and in less development time is the use of multidisciplinary analysis and optimization. Additionally incremental innovation is a driving strategy which intends to incorporate specific solutions for dedicated purposes on existing aircraft models. In this sense the intention of this PhD thesis is to develop a multidisciplinary framework in order to quantify the impact of load alleviation function parameter changes on structural components like the wing and fuselage in terms of resulting mass changes. For this purpose existing in-house Airbus tools and self-developed algorithms are implemented into this framework as a whole. The developed iterative process chain covers the loads calculation including an active load alleviation system, a structural assessment of the wing and fuselage components and a dedicated feedback loop in order to update mass and stiffness properties of the loads calculation model. The structural property optimization of the wing is based on a fully stressed design approach while for the weight penalty estimation of the fuselage section internal loads exceedance values are used. Only vertical balanced manoeuvres and continuous turbulence load conditions are taken into account for this specific use case. The study shows that significant mass reductions are achievable while on the other hand estimated mass penalties are irrelevant.

Kurzfassung

Die stetig wachsende Nachfrage an zivilen Verkehrsflugzeugen durch etablierte Fluggesellschaften in Industrienationen sowie aufsteigenden Wirtschaftsnationen drängt Flugzeughersteller wie Airbus dazu, innovative Lösungen zu entwickeln. Diese Lösungen zielen speziell auf Kostenträger ab, wie z.B. eine Reduzierung der Flugzeuggesamtmasse, ein verbessertes Verhältnis zwischen Auftrieb und Luftwiderstand, eine größere Abflugmasse, einen geringeren Treibstoffverbrauch und dadurch auch auf geringere Kohlendioxidemissionen. Ein Ansatz, um diese Art von Verbesserungen effektiver und innerhalb kürzerer Entwicklungszeiten zu erreichen, ist die Nutzung von multidisziplinärer Analyse und Optimierung. Zudem ist der Ansatz der inkrementellen Innovationen eine wichtige Strategie, welche dazu führen soll, spezifische Lösungen für bestimmte Zwecke an bestehenden Flugzeugmodellen umzusetzen. In diesem Sinne besteht die Intention dieser Dissertation darin, eine multidisziplinäre Analyseumgebung zu entwickeln, um den Einfluss von Parametervariationen des Lastabminderungssystems auf Strukturkomponenten wie den Flügel und Rumpf mit Bezug auf resultierende Änderungen der Massen zu quantifizieren. Zu diesem Zweck werden bestehende Airbus Programme sowie selbst entwickelte Algorithmen in die Analyseumgebung implementiert. Die hierbei entstehende iterative Prozesskette deckt Disziplinen ab, wie z.B. die Lastenberechnung samt Lastabminderungssystem sowie die strukturmechanische Auswertung der Flügel- sowie Rumpfkomponenten. Zudem ist eine Rückkopplung zur Berücksichtigung von Massen- und Steifigkeitsänderungen im Lastenberechnungsmodell implementiert. Die Optimierung der Strukturparameter des Flügels wird mit Hilfe des Konzeptes der voll beanspruchten Struktur durchgeführt, während die Abschätzung der Gewichtszunahme an der Rumpfsektion mittels Enveloppen der Schnittkräfte erfolgt. Für diesen speziellen Anwendungsfall werden lediglich vertikale, ausbalancierte Manöver sowie kontinuierliche Turbulenzen berücksichtigt. Die Studie zeigt, dass eine signifikante Reduzierung der Flügelmasse möglich ist, wobei die geschätzte Massenzunahme am Rumpf vernachlässigbar gering ist.

Contents

List of Figures

List of Tables

List of symbols

Abbreviations

CoL	Center of Lift
AIC	Aerodynamic Influence Coefficient
CFD	Computational Fluid Dynamics
CG	Center of gravity
DLM	Doublet Lattice Method
DM	Dynamic Model
EFCS	Electronic Flight Control System
ELAC	Elevator and Aileron Computer
FC	Forced crippling
FEM	Finite Element Method
FSD	Fully Stressed Design
GLA	Gust Load Alleviation
GPWG	Grid Point Weight Generator
GVT	Ground Vibration Test
HDF	Hierarchical Data Format
HTP	Horizontal Tail Plane
LAF	Load Alleviation Function
LC	Lower Cover
LSF	Load Sharing Facility
MDAO	Multi Disciplinary Analysis and Optimization
MLA	Manoeuvre Load Alleviation
MLW	Maximum Landing Weight
MTOW	Maximum Take Off Weight
MZFW	Maximum Zero Fuel Weight

OWE	Operations Weight Empty
PCL	Patran Command Language
PS	Pure shear
RBE	Rigid Body Element
RF	Reserve Factor
SEC	Spoiler and Elevator Computer
SM	Strength Model
SMT	Shear Moment Torque
UAV	Unmanned Air Vehicle
UC	Upper Cover
ULC	Unitary Load Case
VLM	Vortex Lattice Method
VTP	Vertical Tail Plane

Formula symbols

α^*	Largest real solution of algebraic equation
Δ	Correction value for allowable ultimate shear stress
$\delta_{in,out}$	Inner, outer aileron deflection angle
δ_{SP}	Spoiler deflection angle
η	Normalized wing semispan
ν	Poisson's ratio
ρ	Density
σ	Normal stress value
τ	Shear stress value
τ^*_{allow}	Basic allowable value
ε_m	Mass convergence threshold value
ξ	Internal loads exceedance value
A	Cross section area
A, B, C, D	Coefficient in algebraic equation
c	Chord length

C^*	Material dependent factor
C_l	Lift
C_m	Pitch
D	Hole diameter
E	Young's modulus
F	Force
G	Shear modulus
h	Altitude
I_m	Mass moment of inertia
I_y, I_z	Second moment of inertia around y-axis and around z-axis
J	Torsional moment of inertia
k	Diagonal tension factor
k_t	Net tension efficiency factor
L	Length
$L_{1,...,6}$	Cross sectional dimensions of wing stringer and fuselage frame
M	Pitching moment
m	Mass
Ma	Mach number
n	Load factor
n_{rivet}	Number of rivet lines
q	Dynamic pressure
r	Radius
R_m	Tensile Strength
r_g	Radius of Gyration
$R_{p0.2}$	Yield Strength
S_w	Wing area
t	Thickness value
V_{CAS}	Calibrated Airspeed
V_{mo}	Maximum Operating Speed

Indices

0	Basic coordinates
α	Variable exponent in equation
add	Additional
$allow$	Allowable
cov	cover
e	Elementwise
$g1$	Structural grid points
gg	Global coordinates
$gross$	Gross section
i	Iterator
l, u	Lower, Upper
n	Maximum number of elements / iterations
net	Net section
r	Rotational
t	Translational
tot	Total
tr	Translational and rotational
tu	Ultimate tensile
x, y, z	x-, y-, z-axis

Vectors, matrices and scalars

$\mathbf{f}$	Vector of applied forces and moments
$\mathbf{F}_{bb}$	Flexibility matrix
$\mathbf{G}$	Rigid body transformation matrix
$\mathbf{G}_{12}$	Rigid body transformation matrix from point 1 to 2
$\mathbf{G}_{T2}$	Rigid body transformation matrix from tip to point 2
$\mathbf{K}$	Stiffness matrix
$\mathbf{M}$	Mass matrix
$\mathbf{u}$	Displacement vector

$\mathbf{x}$	Vector of design variables
$\ddot{\mathbf{u}}$	Acceleration vector
x	Design variable
y	Response value

1 Introduction

1.1 The need for a new simulation framework

The global market for civil passenger aircraft is growing steadily. In its global market forecast [Air19] Airbus estimates a rise of so-called *Aviation Mega Cities* from 66 to 95 until the year 2038 with the highest increase in the Asia-Pacific region. Currently these kind of cities, which usually have more than just one airport, account for 40% of the total number of aircraft passengers worldwide. Indeed, the increasing need for passenger aircraft can be explained by the economic growth of emerging markets and the related industrialization and urbanization across such countries. With more people living in cities and their need to travel from one location to another the local airlines have to respond to this demand by increasing their aircraft fleet. During the next 20 years approximately 40,000 aircraft of different types are needed with a market demand of 42% in the Asia-Pacic region (cf. [Air19]). In addition, established airlines in regions such as North-America or Western-Europe are by time faced with the need to replace their existing fleet with new innovative aircraft models due to aging or simply in order to reduce their operating costs and thus increase their profit margin. Another reason is the rise in worldwide tourism and with respect to this the demand for a proportional number of aircraft with different ranges has to be saturated. In this sense aircraft manufacturers like Airbus or Boeing are eager to deliver innovative aircraft models with for instance an improved lift to drag ratio, higher MTOW, less fuel consumption and less carbon dioxide emissions. As stated in [Air19] today's aircraft are already "75% quieter and 80% more fuel efficient per seat" than in the early days of civil aviation.
Innovative approaches by means of numerical simulation can be developed for different phases of the aircraft design process. The aircraft design process consists of the conceptual design, preliminary design and detail design phase. The conceptual design regards the overall aircraft configuration according to requirements and specifications. The preliminary design phase considers the detailed analysis and optimization of parameters from involved disciplines in order to fit the conceptual design. Finally the detail design phase targets a last refinement of structural elements. In terms of development costs late and unplanned modifications of larger scale can cause a high increase in expenses and affect the whole design process negatively. That is why especially the use of numerical simulation is a means of properly setting most of the design parameters in early design phases in order to keep development costs at a low level.
Examples for multidisciplinary problem sets are the optimization of the winglet design for the purpose of less induced drag and thus higher lift at the wing tip region or the update of the flight control system law for the purpose of load alleviation along the wing and thus achieving a structural mass reduction. In order to tackle such kind of problem sets various fields of analysis like the calculation of aerodynamic pressure distribution, the improvement of flight performance using a control system, the preparation of an accurate mass and structural stiffness distribution and the calculation of engine propulsion need to be taken into account. In industrial practice different departments tackle different disci-

plines separated from each other. The common means of interaction between departments is the delivery of results and requests. Using this way of working it can take weeks to receive a feedback for open issues. Another means of handling cross-department tasks and assessing the impact of limited changes on certain disciplines is to take advice from senior experts or use values from technical databases.

The motivation or rather the need for a new multidisciplinary analysis and optimization framework in the industrial aircraft design process arises from these circumstances. Multidisciplinary analysis and optimization has to be deployed in order to get simulation results for interdisciplinary parameter variations in a matter of hours instead of weeks. Especially in the scope of *incremental innovations* which is the working strategy of Airbus in the coming years such a framework makes the quantification of results possible. In this way improvements on the aircraft can be achieved effectively and thus manufacturers like in this case Airbus are able to fulfill the needs of the airlines for profitability in a shorter time frame.

In this sense the aim of this present PhD thesis is to develop a multidisciplinary analysis framework inside the Airbus Flight Physics department. This framework covers those major disciplines sequentially which are gone through during the industry standard development process. The original idea comes from the Component Loads team in Hamburg with the approach to develop a supporting tool for decision making by the chief engineer after preliminary design. The objective is to deploy a first prototype that uses a physically based approach without any surrogate models and in addition shows advancements in aircraft multidisciplinary analysis and optimization. The chosen use case in the frame of this thesis regards the quantification of the impact of load alleviation function parameter variations on the inner/outer stiffened wing box covers and stiffened aft fuselage shell of a generic long range aircraft. Here possible mass reductions along the wing component and related mass penalties for the dedicated fuselage section are quantified based on structural reserve factors and stress values.

Regarding the work at Airbus with an airframe lightweight structure that is getting more complex and engineers who are eager to find an optimal design with minimum weight it is clear that the future is a multidisciplinary one. Engineers need to take into account different disciplines at the same time in order to assess parameter changes more effectively and come to conclusions. In this sense the here developed prototype makes it possible to perform such simulations and to analyze dependencies across disciplines using real simulation models and data in a short time frame. In future studies this prototype can be extended for different types of analysis which can also be located in another phase of the aircraft design process.

1.2 State of the art

An active load alleviation system with varying aileron and spoiler deflection angles changes the lift distribution along the wing such that the center of pressure is located more inboard which yields a reduction of the wing root bending moment and torque. However, reactive elevator deflections are one side effect of this active load alleviation and lead to a force introduction at the aft fuselage section. Related possible mass reductions are calculated via a fully stressed design based structural sizing and mass penalties are estimated using finite element based internal loads exceedance values. Flight manoeuvres and continuous turbulence load conditions are chosen as test cases and the stiffness as well as mass values of the loads calculation model are updated based on the modified structural wing properties. In this section an overview of the state of the art literature and related developments is given.

1.2.1 Multidisciplinary analysis and optimization in aeronautics

In their conference papers [dWASS+07] and [ML13] the authors emphasize that the origins of multidisciplinary optimization can be traced back to structural analysis since in any system an available structure most often interacts with another subsystem. Non-linear programming with its objective function and constraints is among the first approaches to solve a three bar truss problem. Accordingly the mass is minimized with varying geometrical dimensions while stress constraints have to be kept. With a growing number of use cases and maturity the concept of non-linear programming for structural optimization is applied in aeronautics, too. Indeed, when working on structural optimization problems less computational resources make engineers rather work with gradient-free optimality criteria like the fully stressed design approach (cf. [Ven89], [PH98], [PGB93], [RZ91]) instead of time and resource consuming non-linear programming. In his paper [Joh02] E. H. Johnson states that although no mathematical proof is given for the fully stressed design approach to yield a minimum weight design it is still an option in practice to apply it in a first step beforehand any non-linear optimization.
This kind of single discipline optimization has an obvious impact on other disciplines like aerodynamics or performance due to changes along the airframe. As described in [dWASS+07] this respectively leads to an extension of the design space to the related quantities of interest. With respect to existing technical limitations different approaches are suggested to handle these interdisciplinary dependencies. In this sense linear decomposition methods as proposed in [SS82], [WD90], [SJR78], [SS88] and [SSBA83] are brought into practice in order to split the optimization of a larger system into a set of standalone problems, simply to overcome limited computational power. Indeed, this way of decomposition also allows the use of multiple processors in order to speed up calculation time. The idea is to establish an iterative process in which local parameters on subsystem or component level are optimized and wherein these tuned values are then passed to the upper system level for another optimization. In this sense for instance the complete wing box with its stringer cross section area and inertia values is regarded as system level whereas each stringer with its cross sectional dimensions is treated as a subsystem. This linear decomposition approach is applicable to any kind of discipline or aircraft component.
Another factor for the rise of multidisciplinary analysis is the technological advancement in computational resources like more available memory for data storage, the advent of computer clusters or rather super computers, the standardization of parallel processing (cf. [Lub94], [And02]) and the use of high level graphical visualization. This development has two major impacts as described in [MC02] and [dWASS+07]. The first one is the change from low- and medium-fidelity, analytical and database driven approaches and simulation models to high fidelity ones with sophisticated graphics, larger design vector and increased design space. The second impact is a change of the way of working in engineering itself. The trend is towards the use of more simulation models and less physical experiments, mock-ups and technical drawings which then leads to a time and cost reduction of the whole industrial aircraft development process. Considering these aspects multidisciplinary simulation frameworks are enhanced successively for the analysis not just of local structures but of whole systems and rather aircraft configurations.

Different types of analysis sequences are put forth with the rising possibilities of multidisciplinary analysis. In the journal paper [ML13] an overview of different monolithic and distributed architectures with associated ways of model coupling and solutions is given. In the case of monolithic architectures the optimization of different disciplines is performed in just one problem formulation. For instance in his PhD thesis [Jam12] K. James shows the simultaneous topology and aerodynamic shape optimization of an aircraft wing box using the *multidisciplinary feasible* architecture in which "all objective and constraint functions, as well as their sensitivities, are computed in a way that takes aerostructural coupling into account". Both disciplines are coupled by computed forces and displacements. The author states that a better optimization result is achieved when applying this kind of monolithic architecture for his use case (2g manoeuvre case) compared to a sequential optimization of the same model. Especially the benefits of the resulting lift distribution are pointed out with the center of pressure shifted inboard which then leads to less drag due to smaller wing tip deflection and yields a mass reduction, too. Similar approaches are demonstrated in [KM12] and [KKM12]. In the case of distributed architectures each discipline is analyzed and optimized separately whereupon values of design parameters are passed from one discipline to another (cf. [CRR09], [GCdCV08] and [SSF+16]). Simulation models of various disciplines can be coupled based on different fidelity levels. In particular the design of aircraft wings with minimum weight and drag is a widely considered use case. In this sense the authors of [KM12] run simulations of metallic and composite materials for the structural optimization of the wing using a panel aerodynamic model, whereas in [KKM12] the application of high fidelity models for both CFD and structural analysis combined with parallel computing for a highly flexible wing is shown. Aeroelastic tailoring of a composite wing is performed in [DKAG13]. Furthermore reduced order models using various surrogate modeling techniques are investigated in [YM12], [Giu97], [Rum12], [PJB+13], [LHH15] and [SML+16] in order to replace time consuming and intense aerodynamic CFD analysis by cheap function evaluations. Multidisciplinary analysis is an enabler for the investigation of innovative aircraft configurations like BWB (Blended Wing Body) civil airplanes as shown in [HHH08] and [ÖHH01] or other flying wing configurations for military vehicles as investigated in [VK17]. Next to gradient based optimization algorithms also derivative free techniques like *particle swarm* can be used in order to find the global optimum as demonstrated in [VSS04]. Indeed, multiple disciplines can also be combined with algorithms from multi-objective optimization (cf. [KV06]) or multi-body simulation (cf. [Krü08]).
Another trend is the implementation of flight control into multidisciplinary considerations. In this sense R. E. Perez et al. present in their journal paper [PLB06] a developed framework in which flight control during the conceptual design phase is considered. Here they claim that a "sequential process may lead to suboptimal designs" and instead present an optimization framework for simultaneous modifications of the control system, the control surfaces itself and the aircraft conceptual design. Their optimization of a civil passenger aircraft shows an improved design with respect to performance and handling qualities. In [XK11b] and [Xu12] the influence of a manoeuvre load alleviation system and natural laminar flow is studied for aircraft conceptual design. Based on "low-order, physics-based methods" a reduction of direct operating cost and less fuel consumption is achievable. Furthermore in the PhD thesis [Akm06] an active control for the suppression of flutter effects based on an aeroelastic simulation model is investigated. In [ZY07] a transonic aeroservoelastic model for the purpose of active flutter control is developed. In their technical report [BGW99] M. L. Baker et al. discuss mathematical models for aeroservoelastic

analysis at Boeing. In particular they highlight the importance of including "structural dynamic and aeroelastic effects" when designing the control law but in addition urge to keep consistency of quantities which are given in frequency or time domain. In his work [Pal11] N. Paletta studies different aspects of the design, effectiveness, benefits and the quantification of manoeuvre load alleviation systems especially for the purpose of enhanced performance and the extension of structural fatigue life expectations. In their work (cf. [Hag12] and [HML12]) S. Haghighat et al. present an aeroservoelastic framework for the simultaneous optimization of the control system, aerodynamic shape and structural properties of a highly flexible UAV for the purpose of performance improvements. Here they use Euler-Bernoulli beam elements in order to model non-linear wing deflections, apply next to a load alleviation function also model predictive control as well as robust controller and make use of a 3-dimensional panel method in order to analyze maneouvre and gust conditions. Comparable analyses are also demonstrated in [SY93], [POK+19] and [WKP+18] where in [POK+19] the authors emphasize a major intention that "structural weight reduction and high aspect ratio wings play a key role in improving the performance of modern transport aircraft".
It is stated in [dWASS+07] that the integration of such multidisciplinary approaches into an industrial environment plays an important role such that engineering departments in companies can use these new technologies for commercial analysis and development. Such process integration which indeed enables modularity of disciplines is the subject of the next subsection.

1.2.2 Multidisciplinary simulation frameworks in aeronautics

Different frameworks that involve those major disciplines which are tackled in the frame of this thesis are listed in this subsection. In this sense the analysis environment FAME-W is mentioned (cf. [KG97], [KLG95], [VdVKKM00]) which is an Airbus development for quick weight estimation of the wing during preliminary design phase as a result of parameter studies. As noted in [KLG95] this framework is based on "an analytical/numerical algorithm based on the classical theory of multicellular shells, beam and structural instability theory". Indeed, this kind of approach is chosen instead of high fidelity CFD calculations or structural stress analysis based on FEM models in order to enable simulations with less computational resources, too. There is a positive response to this framework especially at the Airbus Future Projects office due to its modularly built structure, the possibility to test new algorithms and available results in a reasonable time frame. Also the impact of an active manoeuvre load alleviation system on the wing bending moment in terms of shear loads and possible wing mass reduction is evaluable including an update of the aerodynamic loads although the calculation of aerodynamic coefficients is based on lifting surface theory.
The optimization framework MBB-Lagrange for structural problems which is developed by the Royal Aerospace Establishment and taken over by Airbus is presented in [BW90] and [Zot92]. The developers of this framework use a FEM model representation of structural components as the central pillar of their optimization approach since detailed structural analysis is more effective using this kind of discretization. In this sense the objective is the minimization of structural weight in which quantities like the element thickness, cross sectional area or the stacking sequence for composite material are used as design variables. It is developed as a modular framework that contains various optimization algorithms and thus makes it possible to connect other software like for the purpose of flutter analysis

and to apply related quantities as boundary conditions. In this way more than only stress boundaries are taken into account. Necessary sensitivity values for the optimization of design variables are obtained analytically due to time reasons but can also be assessed from commercial FEM solvers like MSC Nastran®.
The software ASTROS (cf. [JV88]) is developed for the US Air Force and is supposed to be used during preliminary design for the purpose of time reduction. Also based on a FEM model representation of the aircraft its development is closely based on Nastran notations and file formats. Existing methods for major disciplines are integrated modularly with each module managed and connected using the programming language MAPOL (Matrix Analysis Problem Oriented Language). In this sense the aeroelastic analysis considers steady as well as unsteady aerodynamics and the structural optimization can be performed with the fully stressed design approach or a mathematical optimization including for instance stress, strain or stiffness constraints. Also a static condensation by Guyan is implemented in order to reduce the FEM model for the purpose of modal analysis.
The tool NeoCASS (Next generation Conceptual Aero-Structural Sizing Suite) as described in [CRR09] and [BCDR+08] is based on low and medium fidelity models and especially used for the aircraft conceptual design. It is built in a modular way and combines weight estimation, structural sizing and optimization as well as aeroelastic analysis. The developers put much focus on a reliable weight estimation in early stages including next to the primary structure also the non-structural mass. The process starts with an initial weight estimation based on statistical methods as well as databases and then a regression analysis after the structural sizing is applied in order to update the first estimation. On the other hand no FEM model representation is used for the structural analysis but rather analytical formulas with simplifying assumptions on the structural model. The aeroelastic analysis relies on a beam stick model and aerodynamic panel methods using generalized equations of motion.
P. Piperni et al. from Bombardier Aerospace describe in [PAK04] the integration of in-house developments and other existing tools of various disciplines into a multidisciplinary analysis platform for the purpose of simultaneous optimization of the wing shape and cross sectional properties. Aiming at a reduction of development time during conceptual design phase CFD analysis of different fidelity is implemented and combined with a wing structural analysis as well as weight estimation code. Their idea is to start the optimization with low fidelity models in order to be able to analyze a high number of load cases and then to refine the model in a further step using higher order models. In this sense the software VADOR is used especially for data management and the package EPOGY contains different optimization algorithms. TWSAP is developed in-house in order to generate a wing beam stick model for aeroelastic analysis and in addition to design the wing box which is the basis for their weight estimation module.
In [Gar18] the industrial way of working at Dassault Aviation is described. The major focus of the author lies on the description of the approaches in the field of aeroelastic, aeroservoelastic and aerostructural analysis. The key enablers for these types of FEM model based multidisciplinary studies are the platform ELFINI®, which is used for data management and additionally allows the use of different solvers, and the software CATIA® which is used for pre- and post-processing steps. In standard processes a beam stick model with an accurate stiffness and mass distribution of the aircraft is used in order to be able to calculate thousands of load conditions. Additionally, more sophisticated approaches are developed like the replacement of aerodynamic panel methods and databases for correction by linearized CFD analysis. Central part of the aerostructural analysis is the

global FEM model of the aircraft which is used for internal loads calculation and thus is a basis for the assessment of strength and stability criteria. In terms of aeroservoelastic analysis the author of [Gar18] points out that more research is planned to investigate the use of active load alleviation systems in early development phases in order to achieve mass reductions.

Different frameworks are developed at several institutes of the German Aerospace Center - DLR. For instance A. Schuster et al. from the Institute of Composite Structures and Adaptive Systems present in [SSF+16] their developments regarding the sizing of a civil passenger aircraft focusing mainly on metallic/composite fuselage and wing components. For this purpose they develop a sequential process that covers aeroelastic, aerodynamic and structural analysis. Hereby the generation of simulation models and especially of the global FEM is based on their developed CPACS (Common Parametric Aircraft Configuration Schema) which contains all needed pieces of information like the aircraft mass, center of gravity, position of stringers, frames, ribs, spars and more. The internal loads of the beam stick model after the calculation of load conditions are given in form of Shear, Moment and Torque curves. These values are applied to the global wing and fuselage FEM models in form of discrete forces using rigid body elements without stiffness whose master nodes are placed along the reference axis. The sizing of the fuselage is based on analytical methods first and then followed by numerical simulations. The global FEM static analysis of both fuselage and wing is performed with the ANSYS® solver whereas the sizing is done based on a fully stressed design algorithm using reserve factor values of different sizing criteria until mass convergence.

In their conference paper [KK16] W. R. Krüger and T. Klimmek summarize the developments in the frame of the iLOADS (integrated LOADS analysis at DLR) project by the Institute of Aeroelasticity, Institute of Aerodynamics and Flow Technology, Institute of Structures and Design as well as some other institutes of the German Aerospace Center - DLR. Here the focus is on the loads calculation process and mainly on the aerodynamic analysis as well as on the calculation of manoeuvre and gust loads. The process consists of a definition step of the considered load conditions (e.g. mass, CG, flight points), the loads calculation itself and a post processing in which the resulting Shear, Moment and Torque values are interpreted for load case selection. The model definition and data management is kept in the DLR CPACS, too. Indeed, a number of different DLR developments are used for dedicated tasks. For instance the tool MONA is used for the purpose of creating a parametric FEM model and performing optimization using MSC Nastran®. Another example is the tool VarLoads (which is based on panel methods, a beam stick model and generalized equations of motion) which is used for the loads calculation. The structural sizing is realized with S-BOT or commercial software like HyperSizer® by Collier Research Corporation. Indeed, in the frame of the iLOADS project also different methods for the calculation of aerodynamic loads are investigated like the lifting line theory, 3-dimensional panel methods or Navier-Stokes equations. The resulting loads distribution is applied to the fuselage structure for the purpose of structural analysis (static strength, stability and structural fatigue life). The same approach for structural analysis is also shown in [SSF+16]. Indeed, the authors of [KK16] point out that applying the loads along the fuselage reference axis only might not be accurate enough for sizing purposes. Further studies based on partially the same tool sets are presented in [KDDB+16] in the frame of aeroelastic and aeroservoelastic tailoring of adaptive wing structures and in [DNG12] in the frame of wing mass estimation, including also parts of the secondary structure, using a physical approach.

Different tools for structural optimization are developed at Airbus that can be included into in-house multidisciplinary analysis frameworks. In this sense in [Gri17] the tools PRESTO, ISAMI, ACO and AMO are presented which are all based on a global FEM model of the analyzed components. Essential for this set of tools is the extraction of internal loads and/or sensitivities using the MSC Nastran® solver. PRESTO as a rapid sizing tool is based on a database of discrete internal loads and material properties associated to their related structural reserves. It is developed with the intention to support engineers during early design phases for the purpose of parameter studies. On the other hand ACO and AMO are developed for detailed structural analysis of composite and metallic components applying mathematical optimization. ISAMI is developed to be the official sizing tool at Airbus including a large set of failure criteria. Another development at Airbus regarding the optimization of stiffened fuselage panels is presented in [GSM+09].
The simulation framework PrADO is developed at the Institute of Aircraft Design and Lightweight Structures of the Technische Universität Braunschweig especially for the purpose of aircraft conceptual design studies. Its modularly built structure allows to run different disciplines like aerodynamic or structural analysis in an independent way. Developments on PrADO are coupled with research projects and related PhD thesis. In this sense in [Öst06] the improvement of the framework with a "physical representation of the aircraft configuration" is shown with special attention paid on weight estimation methods and FEM model based analysis. In [Han09] and [HHH08] the extension of the framework to the analysis of a Blended Wing Body configuration is shown. In his thesis [Rie13] the author analyzes composite materials and thus makes use of aeroelastic tailoring. More developments on the framework PrADO are described also in [ÖHH01] and [ÖHH00].
It is worth mentioning that different frameworks are developed by various companies for the purpose of model and tool integration. Examples are given with AML (Adaptive Modeling Language) by Technosoft, CAFFE (Collaborative Application Framework For Engineering), FIDO (Framework for Interdisciplinary Design Optimization) which are developed by the NASA, Isight by Dassault Systems and ModelCenter by Phoenix Integration.

1.3 Objectives and thesis overview

This PhD thesis is the result of a collaboration between the Institute of Aircraft Design and Lightweight Structures of the Technische Universität Braunschweig and the Flight Physics department at Airbus in Hamburg. Its objective is the development of a modular multidisciplinary analysis and optimization framework using a distributed architecture that is directly placed on the Airbus computing infrastructure. A big effort of this thesis is the computational implementation of in-house Airbus tools as well as self-developed algorithms and program code into one framework and thus bring them together as a whole. The process needs to run sequentially with each discipline integrated as a module which means that high fidelity coupled aerostructural or aeroservoelastic analysis are not covered. In addition, parametrization is a core aspect since the framework has to be capable of parameter studies cross disciplines and to assess the impact on multiple aircraft components. Appropriate means, like parallel computing and simulation models of different fidelity, shall be used in order to keep the computation time for a complete analysis loop with hundreds of load conditions reasonable. Furthermore a multi-level decomposition approach for the wing box structural property optimization has to be used in order to implement a fully stressed design algorithm based on static sizing criteria. The idea is to accomplish a first demonstrator based on the latest infrastructure, simulation models and software to lay the foundation for a future standardization of multidisciplinary analysis in a less time consuming aircraft development process.

A physically based approach is applied using global FEM representations of industry standard generic long range aircraft components with proper discretization to yield high accuracy in terms of internal loads distribution and overall stiffness. Only the primary structure of the wing and fuselage are studied and related mass changes assessed as delta values accordingly. Indeed, structural property changes of the wing and subsequent modal analysis with related mass and stiffness variations are taken into account but flutter phenomena, which are studied by many other researchers, are not included.

As a first major use case for this Airbus in-house developed prototype an active load alleviation system as one possible way of mass reduction has to be studied. In this sense the very first target user is the chief engineer after preliminary design who can use the framework as a supporting tool for decision making. With respect to *incremental innovation* the outcome of this PhD thesis has to be a means of quantifying and judging on load alleviation parameter changes regarding their impact on the wing box covers and fuselage stiffened skin panels. Thus the decision of the chief engineer concerning the scale of possible design modifications can be supported quantitatively by results of this simulation framework.

It is worth mentioning that many multidisciplinary analysis and optimization frameworks are developed by different companies or research institutes. However most of them use simplifications in some form, like for instance in the make up of the structural model, apply a small number of load cases, consider a few structural failure criteria, choose a rough discretization of the FEM model or do not use a flight control system. Accordingly many frameworks are used for conceptual studies or preliminary design. Hence the focus of this PhD thesis is to develop a fast approach to achieve reliable and accurate results for a high number of load conditions in a reduced time frame but still under real industrial circumstances. Thus taking into account the major disciplines of aircraft development and combining them in a single framework on high performance servers lays the foundation for more niche use cases like the investigation of incidents during in-service or the analysis of flight test validated loads.

In this sense *chapter 2* contains a description of the developed multidisciplinary analysis and optimization framework. A short explanation of each integrated major module is given with some more exemplary use cases. It follows an overview of the way how the framework is computationally implemented. Some useful post processing routines originated in the frame of this thesis and hence are listed, too.

In *chapter 3* the loads calculation process is described. Here the scope of computed vertical manoeuvres and continuous turbulence conditions with associated mass cases and flight points is outlined. Special attention is paid to the loads calculation model whose accuracy of results, which are provided in form of discrete shear loads over evaluation stations, influences the outcome of the subsequent structural property optimization. The distinct parts of the loads calculation model are listed and detailed, e.g. the applied 3-dimensional aerodynamic panel methods for steady/unsteady calculations, the structural dynamic model which consists of the beam stick model, the mass distribution and the EFCS (Electronic Flight Control System) model. Accordingly the purpose and general aspects of the EFCS are explained with special focus on the manoeuvre load alleviation system and the ways of optimizing its parameters.

In *chapter 4* the structural property optimization and mass penalty estimation are described. This chapter starts with a description of considered material properties as well as modeling aspects of the global FEM representation and its purpose of providing an accurate internal loads flow. Other frameworks, that also make use of the global FEM internal loads, apply the external discrete forces along the loads reference axis of the fuselage only. Contrary to this a *unitary load case* approach is implemented in the frame of this thesis in which the discrete forces and moments are applied along the fuselage according to the physical mass distribution of the secondary structure. The wing box is loaded along the reference axis as it is done in common practice. The results of the linear static analysis are used for both the multi-level wing box property optimization, for which a stress value and reserve factor based fully stressed design approach is implemented, and fuselage mass penalty estimation, that is based on internal loads exceedance values.
The implemented feedback loop in order to pass the updated wing box structural properties back to the loads calculation model is described in *chapter 5*. Here an *equivalent beam* approach is used after the static condensation of the wing box in order to overcome discrepancies in the discretization of both the global FEM *model for strength assessment* and *model for dynamic analysis* which are mandatory used at Airbus. The assessment of delta mass values after the structural property optimization is based on the calculation of concentrated masses with the *Grid Point Weight Generator* whose resulting values are evaluated in the implemented global convergence criteria, too.
Chapter 6 contains the results of the performed design studies. The whole PhD thesis is summarized in *chapter 7* together with a discussion of the applied methods and results. Additional material can be found in the *appendices*.

2 Multidisciplinary analysis and optimization framework

This chapter provides an overview of the developed MDAO process. In this sense its possible use cases are generally described and in addition an overview of the implemented modules, the user interface and data management is given. Also the developed main and post processing routines are explained briefly.

2.1 Developed process and mixture of competences

A global overview of the modularly built MDAO process is shown in the flow chart in Figure 2.1. The developed MDAO process covers different disciplines of aircraft analysis and here especially electronic flight control, manoeuvre and turbulence analysis, stress analysis, structural sizing and a set of algorithms for the feedback of updated structural properties to the loads calculation model are taken into account. Each discipline is implemented as one computation block or rather module for flexibility purposes. In the case of future research this modularity allows the extension of the process chain with more disciplines and further makes the replacement of current algorithms with new ones possible. Accordingly the flight control law or the sizing algorithm can be improved and tested on a targeted aircraft model.

The process is built sequentially but still it is possible to quantitatively evaluate the influence of parameter variations inside one discipline on other disciplines as well as multiple aircraft components. In this sense the main use case that is intended for analysis in the scope of this thesis is the structural assessment of load alleviation function parameter variations for an already sized aircraft model. This scenario is especially important for the chief engineer and accordingly the developed MDAO process is intended to be a supportive tool for better judgment. Other possible use cases are shortly listed as follows:

- Scaling of structural stiffness → impact on results of loads calculation
- Local mass variations → impact on results of loads calculation

For instance property updates of the wing covers result in mass changes which again affect the center of gravity. Further the modifications of the wing structure influence its stiffness which then yields a different bending-torque behavior that has an impact on the angle of attack as well as on the results of the aircraft trim condition. Stiffness changes are also relevant in the scope of continuous turbulence calculations which rely on the related mode shapes.

The MDAO platform consists of self-developed program code and algorithms combined with in-house Airbus tools/software, developments from the Technische Universität Braunschweig and commercial software packages from external suppliers (Matlab® toolboxes, MSC Nastran®). This mixture of competences is detailed in the following.

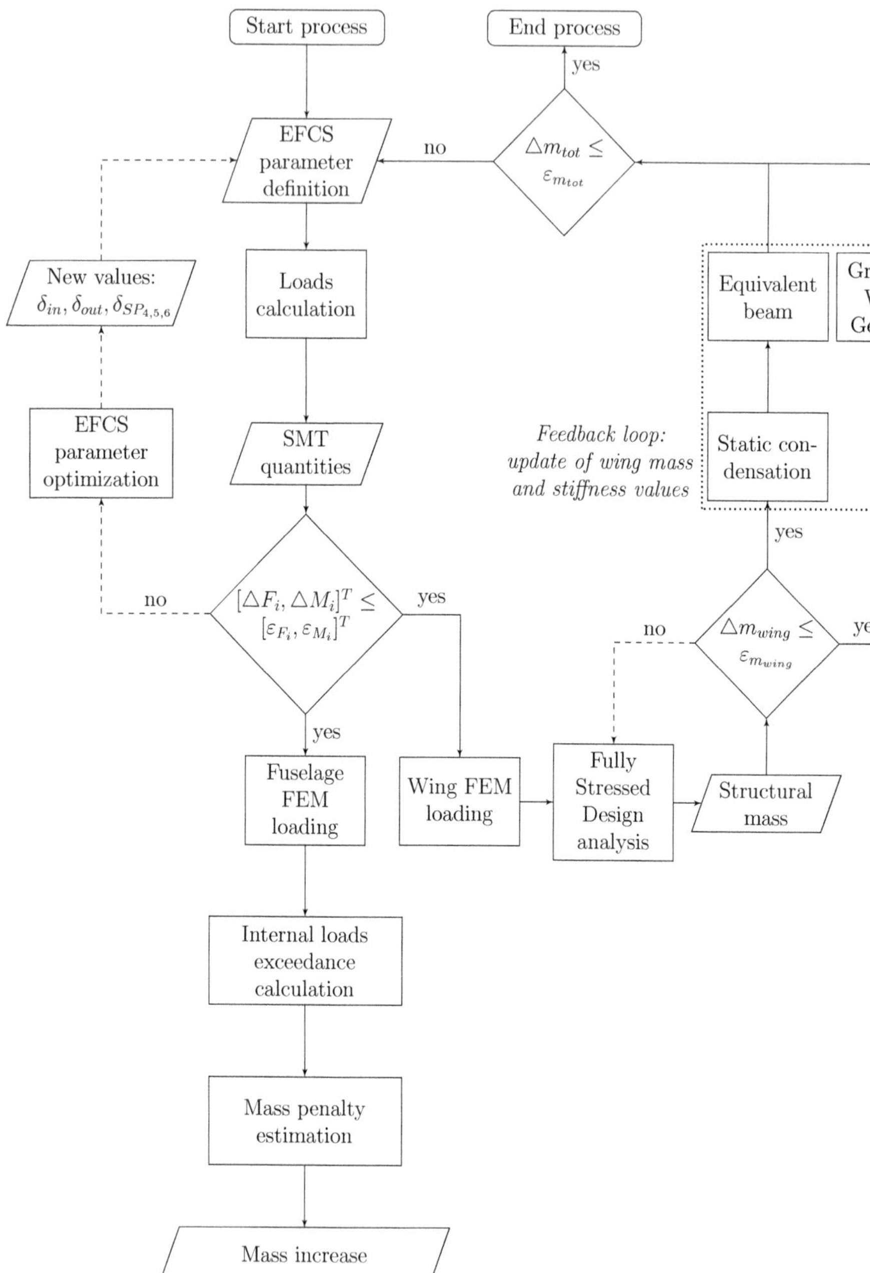

Figure 2.1: Flow chart of global MDAO process

The computational frame of the MDAO platform that steers the input/output data flow, activates modules and runs simulations is self-developed. Also the graphical user interfaces of included tools are skipped in order to run the simulations in batch mode. A complete data set for a long range aircraft is prepared and set-up for the whole process chain. In this sense the following set-up is chosen for the **loads calculation** with EFCS parameter variation:

- Self-developed:
  - ◇ Interface to set and vary EFCS parameters for the purpose of load alleviation
  - ◇ Visualization of results after loads calculation using SMT quantities
- Airbus in-house:
  - ◇ EFCS simulation model
  - ◇ Manoeuvre and turbulence calculation tools
- Commercial software:
  - ◇ Matlab® toolbox for EFCS parameter optimization (not implemented yet)

The set-up of the FEM model based fuselage and wing linear static analysis launching MSC Nastran® is self-developed. The op2 converter to extract the analysis results from the binary file is developed by the Institute of Aircraft Design and Lightweight Structures of the Technische Universität Braunschweig. Beside this the **fuselage** mass penalty estimation process is composed as follows:

- Self-developed:
  - ◇ Set-up of *unitary load case* factorization process in batch mode
  - ◇ Mass penalty estimation
  - ◇ Analysis based on internal loads exceedance values
- Airbus in-house:
  - ◇ Fuselage FEM loading
  - ◇ *Unitary load case* factorization in its core
- Commercial software:
  - ◇ MSC Nastran® GPWG for FEM based mass calculation

The **wing** process for the assessment of a possible mass reduction using a fully stressed design approach is composed as follows:

- Self-developed:
  - ◇ Fully stressed design analysis process
  - ◇ Calculation of possible structural mass reduction
- Airbus in-house:
  - ◇ Wing FEM loading
  - ◇ Structural sizing software for stiffened wing covers

- Commercial software:
  - ◇ MSC Nastran® GPWG for FEM based mass calculation
  - ◇ Matlab® parallel computing toolbox

The **feedback loop** process to send back the updated structural wing properties to the loads calculation model is composed as follows:

- Self-developed:
  - ◇ Connection of feedback loop process frame with previous development steps
  - ◇ Implementation of *equivalent beam* analysis for wing component into feedback loop
  - ◇ Set-up of concentrated mass calculation along wing
- Airbus in-house:
  - ◇ Overall feedback loop process framework
  - ◇ Source code to generate *equivalent beam* model
- Commercial software:
  - ◇ MSC Nastran® for static condensation
  - ◇ MSC Nastran® GPWG for FEM based mass calculation

With the development of the MDAO process several post processing routines originated, which can also be used independently as standalone features for the analysis of generated data. The most important ones are briefly listed here. For the **evaluation of results** various functions are self-written in order to:

- Generate mappings of the wing cover properties
- Generate plots of flight parameter and mass cases

In addition, some routines are written for the purpose of **global sensitivity analysis** based on surrogate models (using Matlab® toolboxes). These are based on data sets that are generated with the MDAO process. The method *extended Fourier amplitude sensitivity test (eFAST)* is chosen for the calculation of sensitivities (cf. [Pet16]).

2.2 General process overview

In the following the global process chain of Figure 2.1 with its main modules is described shortly and additionally each discipline is explained in more detail in the forthcoming chapters. Here the process chain starts with the **definition of the EFCS parameters** which are indeed the related control surface (inner/outer ailerons δ_{in}/δ_{out} and outer spoilers $\delta_{SP_{4,5,6}}$) deflection angles along the wing for the purpose of load alleviation. A simulation model of the EFCS is already part of the loads calculation model which is why solely an interface is developed to steer and transfer the control surface deflection angles into the simulation model.

The process continues with the **calculation of manoeuvre and continuous turbulence cases** which is based on previously mentioned EFCS simulation model and the available loads calculation model with its defined mass and stiffness distributions. An **EFCS parameter optimization** is intended to serve as a technology demonstrator for the determination of optimal control surface deflection angles but is not implemented in the developed version of the MDAO process (dashed lines are chosen in the flow chart to indicate this status). Accordingly the EFCS parameters are varied in the course of a sensitivity study. In the case of an optimization loop the resulting **Shear Moment Torque (SMT) quantities** are used for convergence assessment based on wing and fuselage envelope values. With respect to load alleviation especially forces and moments at the wing root and dedicated aft fuselage cross section are relevant.
For the purpose of FEM model based stress analysis the wing box and the fuselage are taken into account only. Here a linear static analysis is performed and its results are used for the structural assessment. The idea is to quantify a possible mass reduction based on property updates of the inner and outer stiffened wing box covers and to estimate the related mass penalty at the aft fuselage region due to the EFCS parameter variations. In this sense the resulting external loads are applied to a global FEM representation of the considered aircraft components. For the **fuselage FEM loading** a *unitary load case* approach is used which is based on a superposition of factorized discrete forces and moments along the fuselage according to the physical mass distribution of the secondary structure. The fuselage stiffened shell reactive loads after the linear static analysis are then input for the **internal loads exceedance** calculation module. This step compares the internal loads envelopes of a baseline calculation (which is the scenario with no load alleviation) against current values and determines elements of the FEM model that are affected by the new EFCS parameter settings. The **mass penalty estimation** with its resulting **mass increase** value is performed on basis of the number of affected elements and their exceedance values.
In parallel the **wing FEM loading** uses a direct approach in which the discrete forces and moments are applied along the reference axis via rigid body elements. The resulting internal loads are then used for a fully stressed design analysis of the upper and lower stiffened covers. By default the new structural mass value is directly fed into the next module but a switch to multiple iterations until mass convergence is possible (again dashed lines are chosen in the flow chart to indicate this status).
The updated structural properties of the wing box covers are then send back to the loads calculation model via the **feedback loop** module. Note that the FEM model for stress analysis applies to chosen components only but on the other hand the loads calculation model and its related global FEM model (dynamic model, including mass) inside the feedback loop module consider the total aircraft. Figure 2.2 shows the global FEM model representation of a generic long range aircraft model as it is used here inside the feedback loop. Note that a metallic aircraft structure is used through the whole analysis steps of this thesis.
The *feedback loop* starts with a static condensation of both initial and updated wing component in order to use their condensed stiffness matrices in the **equivalent beam** module. Here delta stiffness values are calculated and then transferred to a so-called *tuning bar* which is made of a chain of beam elements along the wing axis of the global (dynamic) FEM model. In parallel the **Grid Point Weight Generator** calculates density based mass values of both wing components and adds the resulting delta mass values to the concentrated masses along the wing axis of the (dynamic) FEM model. If a predefined

mass convergence $\triangle m_{tot} \leq \varepsilon_{m_{tot}}$ for the wing is achieved the process ends otherwise another iteration is started.
Furthermore the developed MDAO platform is applicable for the generation of big data which consists of quantities from different disciplines. For this purpose the whole process chain is run multiple times until the required design space of interest is properly populated. After each module generated data is saved in intermediate files (e.g. as plain text, binary or in Hierarchical Data Format (i.e. HDF5)) which can then be pre-processed for intended data analysis. This data is then processed in order to retrieve new insight or to build surrogate models which are used to replace time consuming process steps and for the purpose of sensitivity analysis. Related examples are shown in [NLHS15a] and [NLHS15b].

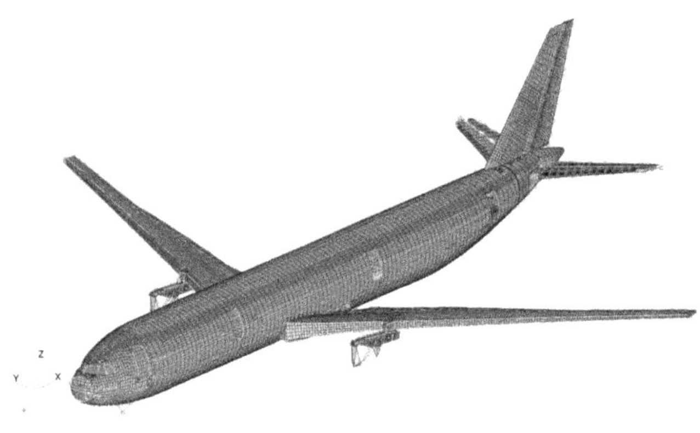

Figure 2.2: Generic long range aircraft model - FEM simulation model

2.3 Aspects of computational implementation

The presented MDAO process is fully developed and implemented in an industrial context with real simulation data and models. The idea is to develop a first prototype for the purpose of feasibility and plausibility in order to proof the general concept. This R&D version of the MDAO process can then be further adapted for the purpose of maturity and industrialization in order to become a standardized software for use inside the company. For the here presented R&D version the chosen main programming language is Matlab® which is used to write and implement own algorithms, select and start modules, steer input/output flows as well as generated data and files. In addition, the programming languages Python and Java® are used for dedicated pre- and post processing functionalities. The combination of these programming languages with Matlab® as the primary choice

allows a quick and more flexible prototyping of this MDAO process. In especially the testing of new ideas using scripting and the use of existing toolboxes as well as available libraries leads to less development time during the prototyping phase. For sure in a later phase when the prototype is going to be industrialized a faster programming language (like the C programming language) is chosen.
The development and data storage is performed on Airbus servers that run with the Red Hat® Enterprise Linux® operating system and which are especially provided by Airbus for time consuming and resource expensive simulations.
In addition, the parallel computing toolbox of Matlab® is used on these servers for iterative algorithms, like for instance the wing box cover optimization, which need to be executed for thousands of finite elements and with different values. The computation is distributed to several cores and in doing so the computation time is reduced according to the number of used cores. Furthermore in order to split and distribute computationally expensive and repeating calculations the platform LSF (Load Sharing Facility) is used. The platform LSF is especially used for compiled software that is already developed like for instance the calculation of manoeuvre and gust conditions as well as the structural sizing of the wing box covers. Here the submitted batch jobs are scheduled according to availability of machines on the servers with the aim to avoid an overloading of the system. The here developed version does not include any graphical user interface but instead the required file paths and switch parameters are passed to the software via a steering file in ASCII text format. Each module has its own data folder and additionally the source code is separated into a main file and files for each module.

3 Loads calculation process

This chapter covers the calculation of the flight manoeuvre and continuous turbulence load conditions inside the MDAO process. This includes a short overview of the main principles, considered flight points, the EFCS and especially its load alleviation function which is one of the main aspects of this thesis. In addition, the design envelopes for the wing and fuselage components are described in terms of SMT values.

3.1 Scope of computed load conditions

The MDAO process considers longitudinal equilibrium flight manoeuvres and continuous turbulence cases but no ground manoeuvres or landing cases. The idea is to develop the first prototype based on these two types of load conditions as a working demonstrator within the limited time frame of this PhD thesis. Indeed, due to its modularity the process can be extended to other load conditions of interest in future developments. For both types of considered load families vertical conditions are taken into account because the LAF (Load Alleviation Function) system applies to these cases only. Hence no lateral cases are analyzed in the scope of this thesis. Further the computed load conditions involve a balanced aircraft model that considers mass, inertia, aerodynamic forces and the required elevator pitching moment. Note that the effect of high lift devices is not included in the analysis. Furthermore no additional cabin pressure or temperature loads are considered in order to limit the number of load conditions although the lack of these cases affects the structural sizing. Hence the structural assessment of LAF parameter changes is performed based on a structural baseline model (cf. chapter 6) and the reference cases, which are outlined in this chapter, only.

In general the temperature loads are applied in order to consider so-called *hot day* and *cold day* cases. These additional loads lead to higher structural deformations and stress values. In addition, the material properties are also affected by means of reduced strength and stiffness values. Further the cabin pressure, that varies with different flight altitudes, increases the tension loads in the fuselage shell and thus provides additional stability to the skin panels. In terms of structural sizing failure criteria this means that a non-pressurized fuselage is more critical to stability criteria. These tension loads due to cabin pressure are also important for fatigue and damage tolerance criteria. In practice the cabin pressure is modeled for the purpose of linear static analysis by applying discrete forces along the fuselage shell which are added to those forces that are generated by the flight manoeuvres and continuous turbulence conditions. However due to the limited time frame the idea is to perform the structural assessment of the fuselage based on internal loads exceedance values (cf. chapter 4.5) only. For this purpose the fuselage mass changes are estimated by comparing the internal loads of the analyzed load cases relative to those ones of the baseline model. Hence the fuselage is not pressurized and fatigue and damage tolerance criteria are not considered, too.

The output of the loads calculation are limit loads. For the purpose of structural analysis

these computed values are scaled by a factor of 1.5 in order to reach ultimate loads level. This change from limit to ultimate loads is performed inside the sizing software based on the analyzed failure criterion.
The equilibrium flight manoeuvres are steady, trimmed and balanced load conditions. Applying this so-called bookcase approach (cf. [WC15]) for symmetric pitching cases the aircraft encounters as described in [WC15] "a steady acceleration normal to the flight path" and in addition the pitch rate is steady as well. The considered manoeuvres are also called pull-up cases for positive load factors and push-down cases for negative ones. For the analysis a six degree of freedom non-linear model is used for which the equations of motion are solved. The equations of motion consider the external forces and also the momentum of the aircraft system based on mass, stiffness and structural damping. In their publication [XK11a] the authors assess the effect of LAF systems on the aircraft performance for manoeuvres and discrete gust cases. For this purpose sensitivity studies and optimizations are performed for a chosen reference design case which is simulated over several time steps in order to assess the aircraft performance. In opposite to the work that is presented in [XK11a] dynamic simulations are not taken into account in this PhD thesis because the main objective here is to assess the effect of the LAF system on the structural mass. In this sense correlated load cases are used which by definition represent the loads acting on the aircraft in a certain time step (cf. [WC15]). In general for correlated load cases the very time step of a dynamic simulation is chosen that yields extremal values. It is also common practice in industry to use correlated loads instead of complete time simulations for stress analysis and the sizing of aircraft components.
For symmetrical manoeuvres steady conditions yield the highest design loads for the wing (cf. [Lom96]). The idea is to play with the LAF system in order to achieve a variation of loads along the wing. These changes in loads consequently allow an adaptation of the wing properties which again results in mass variations. This variation of wing loads and structural properties is restricted because on the other hand in order to keep the aircraft model balanced additional forces are introduced by the elevators which compensate the change in pitching moment due to the wing load variation. These additional forces at the aft fuselage region make an adaptation of the structural properties and thus the mass values necessary (cf. [Lom96], [WC15]).
Next to the flight manoeuvre cases also vertical continuous turbulence load conditions are implemented into the loads calculation step of the MDAO process. Continuous turbulence cases are especially part of the SMT envelope for the wing tip region (cf. [FLW02]). For the computation of the continuous turbulence cases separate steady, balanced and trimmed 1g manoeuvres are simulated first. In a second step positive and negative gust increments are added to the 1g loads in order to have the total load results. Furthermore in contrast to the pure manoeuvre cases the dynamic response of the continuous turbulence is calculated in the frequency domain. Hence for the representation of the aircraft flexibility a superposition of the flexible modes is performed. Note that these flexible modes are retrieved from a free-free modal analysis for which the used mass and stiffness matrices vary with each structural property optimization. The continuous turbulence calculation is based on a stochastic approach which means (as formulated in [WC15]) that a "Gaussian distribution of gust velocity intensities that can be specified in the frequency domain as a power spectral density function" is assumed. Note also that the gust velocity changes randomly, an unsteady aerodynamic model is used and the vertical turbulence causes an additional pitching moment. The most critical turbulence increments are chosen over the considered frequencies assuming the wing root bending moment to be the

dominant selection criterion. The selected increment values are then added to the correlated values of the 1g manoeuvres.
The computed load conditions conform with the Certification Specifications (CS) of the European Aviation Safety Agency (EASA) and also with the Federal Aviation Regulations (FAR) of the Federal Aviation Administration (FAA). In both regulation documents the appropriate paragraphs are written in subpart B (25.255) and C (25.331, 25.333, 25.337) concerning symmetric manoeuvres and again in subpart C (25.341) regarding the continuous turbulence cases. The regulations describing the design airspeeds are listed in paragraph 25.335. In the frame of this thesis special attention is also paid to the effect of speedbrakes on the load distribution along the wing. According to paragraph 25.373 the structure needs to be sized for the described manoeuvres of subpart C with extended speedbrakes as well. The influence of speed control devices is assessed during the calculation of the 1g manoeuvres only when considering continuous turbulence load conditions which is also described in [Lom96]. Note again that the regulations and methodology to compute the desired load conditions are already implemented and available in form of Airbus in-house loads calculation software (cf. section 2.1).

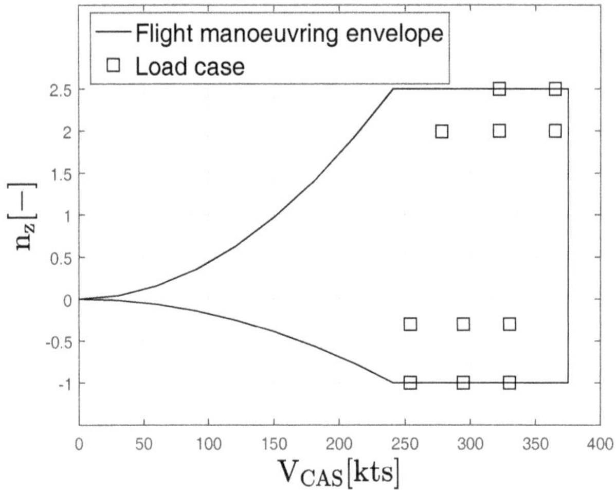

Figure 3.1: V_{CAS} - n_z diagram of applied load cases for design studies

Different design studies are performed using the whole MDAO process chain and a computation time of around 8 hours is targeted for the processing of one total loop. This suggested computation time is based on previous test simulations for the same amount of data, flight points and load cases. In order to achieve the time target with the idea of first prototype computations a limited number of load cases is considered accordingly. This load case set is chosen with respect to the *flight manoeuvring envelope* as described in the regulation paragraph CS 25.333. Here load factors between -1g and 2.5g are chosen using dive and cruise speed which is shown in Figure 3.1. The idea is to combine the envelope load factors together with high speed values. In this sense the figure shows the *flight manoeuvring envelope* together with the selected load cases (each marker represents one

load case) based on their calibrated airspeed against corresponding vertical load factor values. Note that some cases have the same speed and load factor combination but apply a different mass case, speedbrake or thrust configuration. The chosen flight points are based on Airbus templates and in addition more pieces of information about limit manoeuvring load factors and design airspeeds are given in CS 25.337, CS 25.335 and in [Lom96].
In the forthcoming chapter 3.3 the use of different trigger thresholds for the activation of the LAF system is explained. The chosen load case set contains flight points with load factor values below and above the corresponding extreme values (2.5g and -1g), too. This wider range of load factor values allows to study the influence of different trigger values. Note that the different values for the LAF activation trigger need to be set by the user.

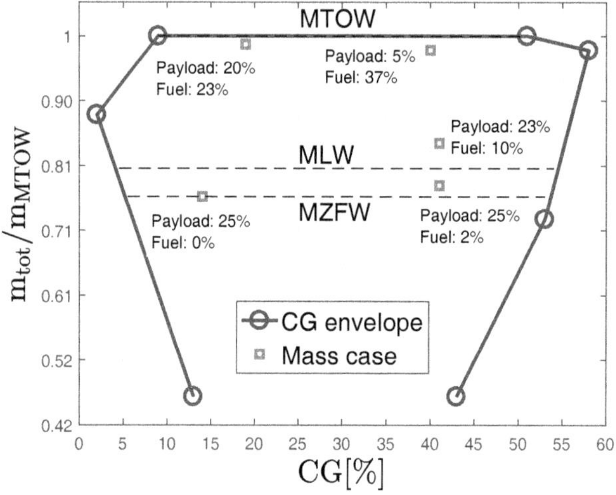

Figure 3.2: Generic long range aircraft weight and balance diagram

Further the load case set considers a certain range of altitudes and Mach numbers (cf. in the appendix, Figure A.1 with altitude values normalized to the highest considered value). The aim is to sweep the available altitude bands and the possible Mach numbers using cruise and dive speed. Here the Mach numbers cover the subsonic and transonic regime. Note that for the analysis of CS-25 aircraft especially the transonic regime is relevant.
In addition, a limited number of mass cases with corresponding center of gravity positions (cf. Figure 3.2 with the total aircraft weight normalized to the MTOW) is also chosen for the generation of the load case set. Here special attention is paid to the partition of the total aircraft mass into payload and fuel. The idea is to cover different mass configurations like MTOW , MZFW and MLW . This partition affects the position of the center of gravity along the fuselage which then influences the intensity of additionally introduced forces at the aft fuselage region due to the pitch compensation. Here again the chosen flight points are based on Airbus templates and in addition more pieces of information about flight altitudes, Mach numbers and mass cases are given in [Lom96] and [Bon09].

Load condition	Number of cases
Vertical manoeuvre	200
Vertical manoeuvre - speedbrakes extended	200
Vertical continuous turbulence	68
Vertical continuous turbulence - speedbrakes extended	68
Total	**536**

Table 3.1: Listing of load conditions for subsequent design studies

Using these certain flight points to compute the load conditions for the subsequent design studies the total number of load cases amounts 536. Table 3.1 lists the number of manoeuvre and continuous turbulence cases. Here a separate listing of cases with extended speed control devices is considered. Further the same table lists maximum and zero thrust cases which are also part of the load case sample. Both thrust modeling scenarios are chosen equally.

3.2 Loads calculation model

The loads calculation model is composed of some core data which are mandatory for the equilibrium manoeuvres and continuous turbulence cases. Here a **structural dynamic model**, also called loads stick model, is used which is built by beam finite elements with the purpose to represent the flexibility of the aircraft. When the stick model is prepared initially its stiffness distribution is varied until the mode shapes and frequencies match reference data from GVT results (cf. [KOCH16], [WC15], [HP11]). An already GVT tuned structural dynamic model is provided by Airbus which is the basis for the wing stiffness modifications due to the structural property changes. This stiffness modification of the structural dynamic model is explained in more detail in chapter 5.
Further a lumped **mass model** is integrated into the loads calculation model. For this purpose concentrated masses are applied to the loads stick model in which each of them has its own inertia tensor and center of gravity. The lumped mass model is used to represent the different CG and mass cases according to Figure 3.2. In general the concentrated mass values can be adapted according to the sphere of interest with regard to fuel loading, the amount of payload and the structural mass. In the case of this work the lumped mass values along the wing need to be updated according to the forthcoming structural property changes. This mass update includes the inertia tensor and the CG position as well which is described in more detail in chapter 5.
The **aerodynamic loads** are calculated by using a panel model of the lifting surfaces. For this purpose aircraft components like the wing, HTP and VTP are discretized with panels that indeed represent also the real geometric dimensions. Figure 3.3 shows some elements of the aeroelastic model of a generic long range aircraft, i.e. the loads stick model (only grid points are shown here) and the aerodynamic panels for the lifting surfaces. Since the grid points of the structural dynamic and aerodynamic model do not coincide an interpolation spline matrix is needed to couple the structural deformation with the aerodynamic forces (cf. [RJ04], [WC15]). Here in a first step the structural deformations are interpolated to the aerodynamic ones and secondly as written in [RJ04] a transformation is performed that yields "the relationship between the aerodynamic forces

and the structurally equivalent forces acting on the structural grid points". In the appendix Figure A.2 shows how the splining connects grid points of both disciplines inside the aeroelastic model. In Figure A.3 the complete aeroelastic model can be seen.

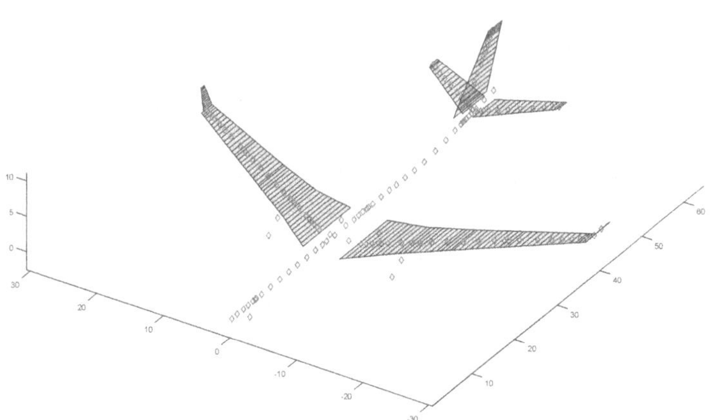

Figure 3.3: Generic long range aircraft aeroelastic model: structural grids + aero panels

The calculation of aerodynamic loads using 3-dimensional panel methods is based on potential flow theory for which either vortices or doublets are used. Here linear aerodynamic theory is applied which assumes a small angle of attack and both incompressible as well as subsonic flow (cf. [Mel00]). Vortices are the basis for the so-called Vortex Lattice Method (VLM) which is capable of calculating quasi-steady aerodynamic loads and is applied for the computation of equilibrium manoeuvres. The term quasi-steady suggests that the lift reaches its steady value instantly whereas on the other hand when assuming unsteady aerodynamics the lift reaches its steady value asymptotically (cf. [WC15]). One important aspect of aerodynamic panel methods opposite to 2-dimensional strip theory is the interaction of aerodynamic forces, which are modelled using horseshoe vortices derived from Biot-Savart's law, along neighbouring panels. This interaction effect is covered in the so-called Aerodynamic Influence Coefficient (AIC) matrix. For linear quasi-steady aerodynamics the AIC matrix consists of real parts only mathematically speaking and the so-called reduced frequency, since no frequency dependent effects are considered (cf. [VCO17]), is equal to zero. In his thesis [Mel00] the author describes the main aspects of the VLM and also provides a Matlab® implementation that is available for interested developers. In the technical report [Hed66] S. G. Hedman describes the theoretical basis of the VLM and presents a study regarding the influence of different panel patterns on the calculated lift coefficients.

For the calculation of continuous turbulence cases linear unsteady aerodynamics are applied. For this purpose the VLM is extended to the Doublet Lattice Method (DLM) that uses as stated in [WC15] for each panel an additional "acceleration potential doublet"

next to the horseshoe vortex for the modeling of the aerodynamic forces during unsteady motion. Thus the AIC matrix consists of real and imaginary parts and is also a function of the reduced frequency. Here the unsteady contributions are added to the steady ones in the time domain by using a rational function approximation (cf. [VCO17], [WC15], [Res06]). A general derivation of the DLM and its mathematics is described in the article [AR69] and in the technical report [Bla92] whereas the book [KP91] provides a broader overview about wing theory and panel methods. In his report [Mar78] the author summarizes paneling techniques for different aircraft components and also discusses the effect of trailing vortices on downstream panel elements. The here used DLM method follows the MSC Nastran® implementation (cf. [RJ04]) for subsonic flow and slender bodies and is already part of the Airbus software for loads calculation.
Using 3-dimensional panel methods the computed forces can be integrated to calculate different aerodynamic coefficients and stability derivatives (cf. [Mel00], [VCO17], [RJ04]) which are again used to assess the dynamic behavior of the aircraft. Note that panel methods imply induced drag only and viscous drag is not covered. One idea to consider viscous effects when using panel methods is to superimpose lift and drag coefficients using a drag polar as performed in [JPM10]. Alternatively this kind of viscous effects can be accomplished by CFD analysis.
In general aerodynamic panel methods are computationally inexpensive but on the other hand less accurate than high fidelity CFD calculations, i.e. when lift distributions for the transonic regime are calculated. As stated in [BSB00] the inaccuracy of VLM and DLM is based on several idealizations like for instance the exclusion of non-linear effects as well as viscous effects and the assumption of thin lifting surfaces. The transonic flight regime plays an important role for the analysis of CS-25 aircraft. Hence in order to compute the lift distributions for a high number of load conditions in a short time frame the panel methods are chosen although the results are less accurate. A reasonable compromise is to extend panel methods with an additional correction of the AIC matrix based on a least squares method and data from flight tests, wind tunnel tests as well as selectively chosen CFD calculations for different flight configurations (angle of attack, sideslip angle, Mach number, etc.). Note that the data provided from the wind tunnel tests are for steady states and rigid aircraft only but using the complete **aerodynamic database for correction** also non-linear effects are taken into account. The authors of the conference paper [BSB00] describe the "least squares AIC correction method" which is used inside the MDAO process and also list other approaches for AIC corrections. In other examples (cf. [KFN17] and [TD14]) a quasi-steady CFD sample based AIC matrix correction method, that relies on a Taylor series expansion, is demonstrated. It is worth mentioning that intense research is performed to achieve results from CFD calculations in a shorter time frame in order to be able to use them directly for studies in the transonic regime. For example the authors of [KFN17] and [WSTA+17] compare in their conference papers the results of a time-linearized CFD calculation against non-linear CFD and corrected DLM results.
In the CS 25.301 it is prescribed that "if deflections under load would significantly change the distribution of external or internal loads, this redistribution must be taken into account" and according to this specification effects of a flexible structure are also taken into account in the loads calculation of the MDAO process. The adjustment of aerodynamic forces and moments due to this effect of **structural flexibility** is based on corrected panel methods and performed by applying so-called flexible factors (cf. [RK07] and [BZA87]), which indeed depend on the Mach number, dynamic pressure and on the

aerodynamic coefficients of the total aircraft. These increments of aerodynamic forces and moments are retrieved from the aerodynamic database using different flight parameters, the aircraft geometry definition and the stiffness distribution as inputs. In order to achieve the load distribution of a flexible structure these increments are superimposed during the simulation of the load case. Here the coupled wing bending/torsion deflection and the resulting change in the angle of attack are the most significant aspects that are considered. Figure 3.4 shows the effect of a flexible wing on the normalized lift distribution along the normalized wing semispan η for a 2.5g pull-up manoeuvre. It can be seen that structural flexibility has a significant impact on the lift distribution and accordingly on other quantities, too. Note that in this way the lift decrease at the wing tip due to structural flexibility is taken into account as well. Using this approach based on the correction/adaptation of aerodynamic coefficients one speaks of quasi-flexible aerodynamics (cf. [WC15]). The structural property optimization of the wing covers and the resulting stiffness changes are important aspects which indeed have an effect on the results of this flexibilization process.

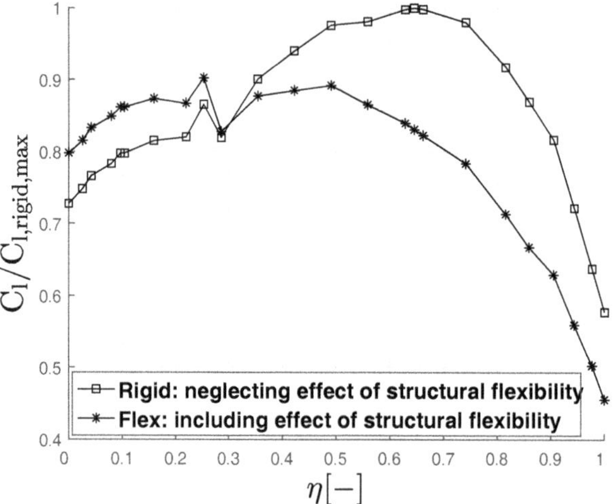

Figure 3.4: Effect of wing structural flexibility on normalized lift distribution: 2.5g pull-up manoeuvre

Further a simulation model of the **EFCS** is also part of the loads calculation model. Changes that are performed during the structural property optimization affect the wing stiffness and mass distribution of the loads calculation model and thus the aircraft loads indirectly. On the other hand the EFCS has a direct influence on the loads. In general the purpose of the EFCS is to control the aircraft longitudinal and lateral movement by deflecting its movable control surfaces, e.g. ailerons, spoilers, slats, flaps, rudder and elevator. The control surfaces of a generic long range aircraft are shown in Figure 3.5.
In addition, LAF parameter variation for the purpose of active load alleviation, which is part of the EFCS, is applied to alleviate the loads along the wing for vertical conditions by aileron/spoiler deflections. Here an interface is used to steer the aileron/spoiler deflection

angles for the purpose of load alleviation during trim conditions and transfer these values into the simulation model of the EFCS. The EFCS is described in more detail in the forthcoming section.
Further **geometry definition** and **design case definition** files are used. The former is used in order to associate structural grid points with aircraft components and the latter to define load case families and flight points.

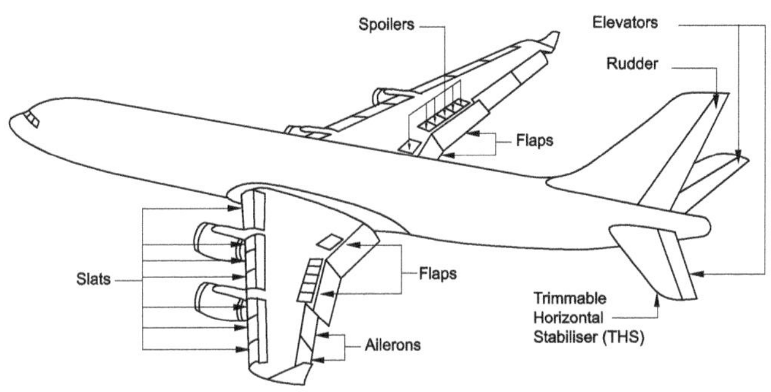

Figure 3.5: Generic long range aircraft control surfaces according to [Air00]

The **aircraft reactive loads** are calculated using the 1-dimensional structural dynamic loads stick model which allows to retrieve the values of interesting quantities at discrete stations. These retrieved values vary based on the considered manoeuvre and gust conditions which lead to deformations and thus reactions of different intensity. The so-called force summation method (cf. [WC15]) is applied in order to calculate the values of interesting quantities at dedicated cross sections of the finite beam model. Note that evaluation points for aerodynamic and mass values can be located differently. Since the most interesting quantities are the shear, moment and torque values the reactive loads are also abbreviated as SMT values. When plotting SMT values along different aircraft components the resulting curves can for instance be used for the sorting and selection of critical load conditions. Figure 3.6 shows part of the loads stick model and schematic SMT curves of the vertical bending moment for the wing and fuselage components. In practice SMT extremal values, like for instance the maximum vertical bending moment at the wing root, over all considered load cases are used in order to select the critical load cases. Here SMT values are especially used to assess the influence of LAF parameter changes on the wing and fuselage structure.
In general the reactive loads are the result of aerodynamic, inertia and additionally applied discrete forces. As mentioned before the aerodynamic forces are calculated using a panel method and the inertia forces are related to the discrete mass distribution. The idea of externally applied discrete forces is meant to consider especially the effect of engine mass and thrust as well as landing gear effects in the case of ground conditions. Since

the wings are swept the engine thrust and mass result in a discrete change in the shear, moment and torque value at the engine installation point.

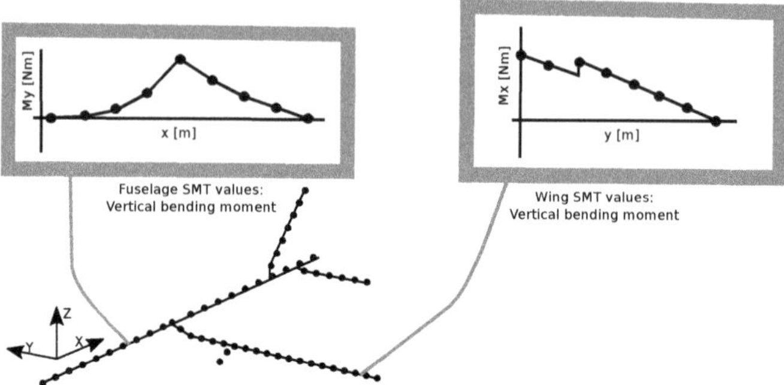

Figure 3.6: Part of 1-dimensional loads stick model and schematic wing/fuselage SMT curves of vertical bending moment

3.3 Electronic Flight Control System

3.3.1 General aspects of the Electronic Flight Control System

The EFCS is used to control the movements of the aircraft during flight and landing by its movable lifting surfaces. As described in [ZD84] and [BAL11] in modern long range aircraft, as it is the case in this thesis, a fly-by-wire EFCS is used. Opposite to flight control systems, which rely on hydro-mechanical designs, there is no mechanical connection between the pilot command elements and the lifting surfaces anymore. Instead the pilot command is transferred through a fully electric control system based on digital technology to hydraulic actuators which then deflect the control surfaces. Figure 3.7 shows the general architecture of a fly-by-wire system for aileron and spoiler deflection. Here the computing units are the ELAC (Elevator and Aileron Computer) and the SEC (Spoiler and Elevator Computer) . Note that the autopilot is a separate unit. By default the aircraft flies in *"normal law"* and here the *"pitch control"* is the most relevant part of it for the computed vertical equilibrium manoeuvres and turbulence cases. The *"normal law"* includes in particular a flight envelope protection and the MLA (cf. [Air]). The MLA is described further in the next section. Another advantage of the fly-by-wire EFCS (which is not possible with hydro-mechanical flight control systems) is the mass reduction due to the lack of mechanical components. Further advantages are listed in the book [BAL11] and the conference paper [ZD84].

The EFCS for *"pitch control"* works in *"normal law"* in the case of no system calamities and is then driven by the vertical load factor n_z that is measured by sensors like accelerometers. The *"normal law"* changes to *"alternate law"* when a certain number of system calamities appear and in even worse cases becomes a *"direct law"*. This downgrading of the control law is equivalent to an increasing loss of transmission of the pilot command to the control surface actuators by electric signals. For the worst case sce-

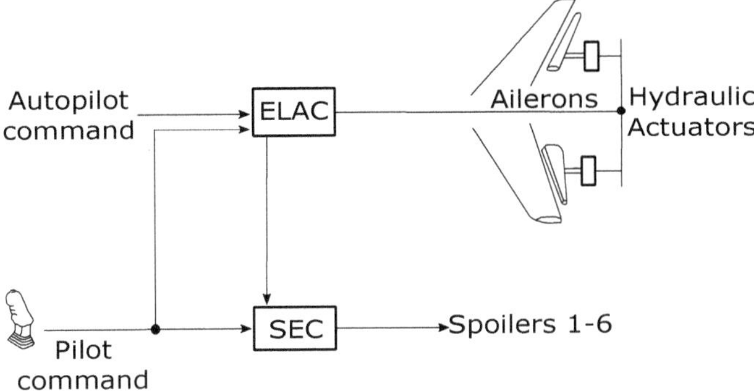

Figure 3.7: General architecture of fly-by-wire EFCS for aileron/spoiler deflection based on [BAL11]

nario it is mentioned in [ZD84] that a "minimum mechanical back-up will be ensured by a mechanically controlled rudder and standby trimable horizontal stabilizer control" in order to land the aircraft safely. Note that as further safety precautions EFCS units are installed redundant and diversified (in terms of hardware and software) with the aim to reduce the risk of total system failure. Furthermore special for the EFCS is the flight envelope protection with the help of different laws. Here as stated in [ZD84] and [Air] most mentionable beside the load factor limitation are the protection laws for overspeed, high angle of attack and pitch attitude.

3.3.2 Manoeuvre load alleviation

It is possible to alleviate the loads along the wing in different ways, i.e. by an active load alleviation system or by aeroelastic tailoring for composite material and also by the variation of the fuel distribution to achieve inertia relief. The focus here is on the active load alleviation. As mentioned before the MLA is part of the EFCS *"normal law"* and its purpose is to reduce the loads along the wing in the case of vertical load conditions. In the conference paper [GB84] the authors mention two main advantages through the use of the MLA which are an expected wing mass reduction and based on this a decrease of direct operating costs. The qualitative change of lift due to an active load alleviation for positive load factors can be seen in Figure 3.8. For the purpose of load alleviation the vertical load factor needs to surpass a predefined threshold value in order for the ailerons and spoilers to be deflected symmetrically upwards. The generated lift depends on the shape of the airfoil and in this sense the shape of the airfoil at the outer wing region is changed by its aileron and spoiler deflections. Accordingly an upward deflection results in a decrease of the generated lift at regions along the wing span where these movable lifting surfaces are installed. The total lift needs to stay the same (with and without the MLA) which means a redistribution of the lift along the wing span and thus a shift of the center of pressure inboard, too. Indeed, a reduction of lift at the outer wing region also means a reduction of the wing root bending moment and torque. For the purpose of this thesis a negative load factor scenario with an active load alleviation is explored as well. In this case only the ailerons are deflected downwards since spoiler deflections are inhibited in this direction. The center of pressure is shifted outboard accordingly and thus the negative bending of the wing is reduced. An active MLA is connected with an additional pitching moment originating by aileron and spoiler deflections (cf. [Air]) which needs to be compensated with an elevator demand. Effectively on one hand the MLA leads to a reduction of loads along the wing but on the other hand additional forces are introduced by the elevators, which compensate the change in pitching moment, at the aft fuselage region. These opposing load changes are a core aspect of investigation in this thesis. A short overview of the design and realization of an active load alleviation system is given in the paper [GB84].

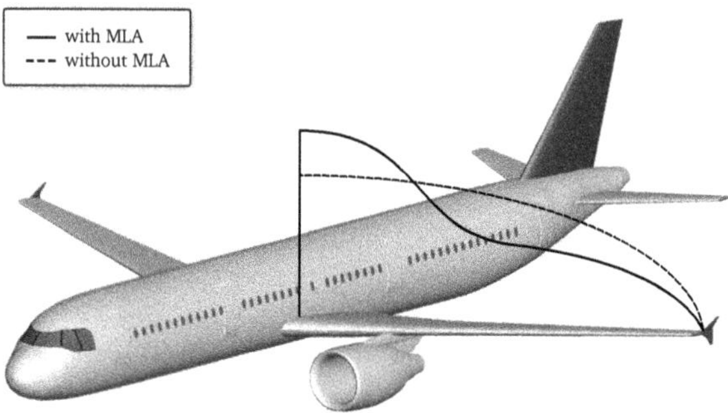

Figure 3.8: Qualitative change of lift distribution with and without MLA, taken from [NLHS15a]

As mentioned before the computed equilibrium manoeuvres are steady, trimmed and balanced conditions. For these cases time simulations are not performed and hence for the simulation rather limits for the flight control law are imposed than applying the full EFCS (cf. [WC15]). On the other hand, if time simulations have to be computed the full EFCS simulation model still can be used. Using these limit values for the various control surface deflection angles the idea is to further provide aircraft stability as well as to stay in line with handling quality targets. Handling qualities concern the requirements regarding the manoeuvrability of an aircraft by the pilot during climb, cruise, landing and further conditions. This discipline involves the quality of pilot command processing through the different control laws. With respect to [BAL11] the dynamic behavior of the aircraft after the pilot command as well as navigating along the targeted flight path accurately play an important role for the assessment of handling qualities. Further the authors of [BAL11] highlight the necessity to apply handling quality targets such that during all considered flight missions the pilot command does not lead to additional oscillations due to aircraft instability. Note that one option to assess the quality of the control system and its handling qualities is to evaluate the judgement of the pilot (cf. [BAL11], [BJSH66], [USA86]). In their technical note [BJSH66] the authors present an evaluation program for qualitative and quantitative assessment of handling qualities based on the judgement of the pilot. Here they especially use stability and control characteristics. Further they highlight more reasons for instability effects and worsening handling qualities like for instance a center of gravity that is located more aft or extended landing gears and flaps. In a similar way using pilot opinion the authors of [USA86] refer in their technical report to the so-called Cooper-Harper pilot rating scale for the evaluation of handling qualities. Note, as stated

in [BAL11], in general the development of a control law needs to take into account criteria like aircraft stability and damping, manoeuvrability, passenger comfort, flight path accuracy, low operating costs and flight envelope protection. Although handling qualities are an important aspect which needs to be considered the effect of MLA parameter variation on handling qualities is not further studied in this thesis.
For the continuous turbulence cases the approach is to compute the 1g equilibrium manoeuvres first and then to add the corresponding turbulence increments to the results. For the 1g manoeuvres also control law limit values are applied instead of using the full EFCS. In their patent document [KG00] R. Kelm and M. Grabietz describe a method to reduce total wind gust loads by prophylactic reduction of the 1g loads. The same principle is used in the frame of this thesis. The idea is to alleviate the loads along the wing already for the 1g trim before the occurrence of the turbulence. In [KG00] R. Kelm and M. Grabietz highlight that on the one hand additional aerodynamic drag is introduced due to the aileron and spoiler deflections but on the other hand the wing structural sizing can be adapted according to the alleviated loads and hence one can consider a reduction of the total aircraft mass which again yields less fuel consumption. Note that less fuel consumption is a strong argument in favour of an active MLA system in particular for long range aircraft. Although the focus of this thesis is on the MLA system it is worth mentioning that a linearized GLA model is used since the continuous turbulence solution is performed in the frequency domain as stated in [WC15]. In principle the load alleviation system for gust has the same objectives as the one for manoeuvre cases but as highlighted in [WC15] it acts faster when it comes to shift the center of pressure inboard. While the MLA parameters are modified here the GLA ones are used as provided.
In this thesis for the purpose of MLA parameter variation an interface is coded using Matlab® (cf. section 2.3 regarding aspects of computational implementation) which in the end transfers the target aileron and spoiler values to the simulation model of the EFCS. The workflow of this interface is explained as follows. The user defines those mass cases (as explained in section 3.1), calibrated airspeeds and vertical load factors which are all indeed an input for the loads calculation process. In a first step (cf. equation 3.1) the interface checks if the mass condition m_{tot} exceeds a prescribed mass value $m_{tot,trigger}$ in which case the load alleviation system is activated. By default the mass scheduling is switched off in the scope of this thesis but it is implemented into the process chain for future research.

$$m_{tot} > m_{tot,trigger} \tag{3.1}$$

An additional trigger (cf. equation 3.2) based on the calibrated airspeed V_{CAS} is implemented as well in order to consider the load alleviation for high speed cases only.

$$V_{CAS} > V_{CAS,trigger} \tag{3.2}$$

Composed of the pieces of information given in [BAL11], [Air] and [Air00] the range for the speed trigger is chosen as shown in equation 3.3. Here the upper boundary is set to the maximum operating speed V_{mo} with an additional tolerance value of $10kts$.

$$250kts < V_{CAS} < 340kts \tag{3.3}$$

Actually the most important trigger (cf. equation 3.4) is the vertical load factor n_z for which here absolute values are used so that both positive and negative scenarios can be studied using the same interface. In the case the mass or the speed trigger are not exceeded the MLA is switched off completely for the corresponding load condition. Otherwise for positive load factors the usual trigger is set at $n_z = 2.0g$ (cf. [Air]) and for negative ones at $n_z = -0.3g$ (Airbus recommendation for this thesis).

$$|n_z| > |n_{z,trigger}| \tag{3.4}$$

Using an active load alleviation the deflections of the aileron and spoiler control surfaces are chosen directly proportional to the vertical load factor (cf. equation 3.5). Starting from the load factor trigger value both ailerons and spoilers are deflected linearly until their maximum defined deflection angles are reached. The target deflection angles for the inner and outer ailerons ($\delta_{in}, \delta_{out}$) and the outer spoilers ($\delta_{SP_{4,5,6}}$) are in the end passed to the simulation model of the EFCS. These steps of this described EFCS interface are performed for both the equilibrium manoeuvres and the 1g trim conditions which are needed for the turbulence cases. Note that the mass as well as the load factor trigger can be different for the two load case families.

$$[\delta_{in}, \delta_{out}, \delta_{SP_{4,5,6}}]^T \sim n_z \tag{3.5}$$

Technically the ailerons of a long range aircraft like for instance the A340-600 can be deflected upwards and downwards by 25°(cf. [Sad12]). But indeed a set of recommended MLA parameter settings are compiled according to different sources which are listed in the following. Next to the patent document [KG00] and the widely used book about aircraft control [BAL11] further two Airbus training documents for pilots (cf. [Air00], [Air]) are used as references for the proposed deflection angles. The compiled MLA parameter settings which are used for an initial design study later on are shown in Figure 3.9.

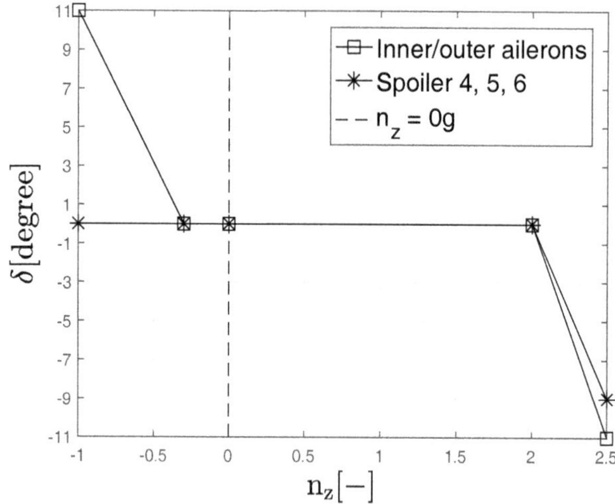

Figure 3.9: Recommended aileron/spoiler settings for initial design study of MLA

Here the trigger values for the vertical load factor n_z and the fact that spoilers can be deflected upwards only are considered. In this sense negative values indicate upward

	δ_{SP_1}	δ_{SP_2}	δ_{SP_3}	δ_{SP_4}	δ_{SP_5}	δ_{SP_6}	δ_{in}	δ_{out}
1g LAF	0°	0°	0°	0°	0°	0°	-11°	-11°
Speedbrakes extended	-25°	-30°	-30°	-30°	-30°	-30°	0°	0°

Table 3.2: Recommended aileron/spoiler settings for initial design study of 1g LAF and load cases with extended speedbrakes

deflections and positive values downward deflections of the control surfaces. As mentioned before the ailerons and spoilers are deflected linearly after the trigger threshold value is passed. For example the MLA is activated for a 2.5g pull-up manoeuvre with the inner and outer ailerons deflected -11°upwards and the spoilers 4, 5, 6 deflected -9°upwards.
The initial settings for the LAF of the 1g trim conditions and for the spoiler deflections of those load cases with extended speedbrakes are listed in Table 3.2. Note that the speedbrakes are extended for load cases with vertical load factor values higher than zero only. Furthermore the 1g cases do not require a load factor trigger like the one that is applied for the equilibrium manoeuvres.
Further note that in the control law the functionality of load alleviation is put on higher priority than the one of the speedbrakes. This means that for a manoeuvre with extended speedbrakes using the spoilers $\delta_{SP_{1,\ldots,6}}$ an active load alleviation system decreases the deflection angles of the spoilers $\delta_{SP_{4,5,6}}$ to those values of the prescribed MLA. In this way the high loads at the mid and outer span of the wing that appear due to the extended speedbrakes are reduced (cf. [FLW02]). Another effect one has to keep in mind when working with load alleviation systems is the possible change of the order of critical load conditions. For instance due to the different triggers which are used those affected load conditions become less critical but on the other hand those excluded load conditions can become relevant for the sizing of the structure.

3.3.3 EFCS parameter optimization

The most desirable scenario is a full EFCS optimization with all its constraints and objectives. However as mentioned before in the frame of this thesis a full EFCS optimization is not performed because correlated cases (cf. section 3.1) are considered only. In addition, an adequate level of maturity is not reached in order to perform the originally intended multi-objective EFCS *parameter* optimization and hence this step is a certain topic for future research. Therefore an EFCS *parameter* variation is performed which is based on the recommendations that are shown in the previous section 3.3.2. The impact of this EFCS *parameter* variation on the aircraft structure is analyzed using the complete chain of the sequential MDAO process. The results of this analysis are shown in chapter 6.
In general the first step before an optimization consists of a sensitivity analysis in order to determine those parameters that have the highest influence on the objective function. In this sense the outcome of a sensitivity analysis of EFCS *parameters* based on part (from loads calculation to wing structural analysis only) of the here developed MDAO process is shown in the conference paper [NLHS15a]. The sensitivities of manoeuvre load alleviation parameters on buckling reserve factors using surrogate model based *"extended Fourier amplitude sensitivity test"* are assessed. The method *"extended Fourier amplitude sensitivity test"* is described in [NLHS15a] as a "variance decomposition method for global sensitivity analysis". The central ideas are described in the article [STC99] and a computational implementation is shown in [MTS82]. As stated in [NLHS15a] "the use of

surrogate modeling techniques helps to avoid the time consuming high fidelity analysis but still provides accurate results. In the frame of global sensitivity analysis the surrogate model is used to apply variance based *"extended Fourier amplitude sensitivity test"*. The contribution of input parameter variation of the manoeuvre load alleviation system to the variance of the surrogate model output in terms of buckling reserve factors is measured. The sensitivity study is performed on the upper cover of a backward swept composite wing and the results are compared to those of the high fidelity analysis. Note that the variation of maneuver load alleviation parameters is nowadays assessed by external loads resulting from the flight maneuver calculation only. The presented approach includes reserve factors and hence provides an insight into the structural response. In this way those maneuver load alleviation parameters are found that affect the structure in terms of buckling reserve factors the most and can be used for future design changes and weight reductions". The surrogate model is built using a method called *high dimensional approximation* which is very similar to artificial neural networks and is based on a superposition of different transfer functions like sigmoid, radial basis and linear ones. The global sensitivity analysis method called *"extended Fourier amplitude sensitivity test"* is applied to this surrogate model, which relates MLA and other flight parameters to buckling reserve factors of the upper cover wing, in order to determine the most influential loads parameter on the structural response. The significance of the aileron deflection angles on the buckling reserve factor is shown.

Further a structural capability analysis is performed by F. B. Peters and its outcome is documented in his Master Thesis [Pet16]. Here again both the developed MDAO process and the surrogate model based *"extended Fourier amplitude sensitivity test"* are used. The same approach as shown in [NLHS15a] is used in order to determine the most important SMT quantities along the wing evaluation stations with respect to the minimum structural reserve factor. The dominance of the vertical wing bending moment on the strength of the upper and lower wing covers is shown in a quantitative way. In addition, the direct influence of aileron and spoiler deflections on the load variation and the indirect influence of the same control surfaces on the reserve factors are shown, too. In this sense the chosen SMT quantities are then scaled up locally (at the region of highest influence) such that a reduction of the reserve factor is achieved in order to assess the possible margin until structural failure.

In order to be able to consider a full EFCS optimization the load cases need to be simulated over several time steps. However the main focus here is put on the MLA of the EFCS which is used during the computation of the equilibrium manoeuvres and the 1g trim conditions. Since correlated load cases are considered only the possible scope of EFCS optimization is limited. In this sense one idea for an optimization is to tune the previously mentioned MLA parameters (aileron/spoiler deflection angles) with regard to the resulting SMT values at the wing root and aft fuselage section. In this case the possible optimization approach is rather called an EFCS *parameter* optimization. The idea for this optimization approach is outlined as follows.

The MLA parameter variation results in a redistribution of the lift C_l along the wing span which is shown in Figure 3.10 for a 2.0g equilibrium manoeuvre (in this example plot the MLA n_z trigger is different than the one that is shown in Figure 3.9). The lift values are normalized here. In this example case the inner and outer ailerons are deflected upwards by -10°which yields a shift of the center of lift (shown in the figure as ΔCoL) inboard. Indeed, next to the visualization of the lift distribution another important implementation into the MDAO process is the quantification of the center of lift and its alteration in accordance to LAF parameter changes. This information can also support the judgement concerning the LAF impact since the location of the center of pressure is the main driver. Note that throughout the thesis the assumption is made that an inactive LAF is equivalent to 0°aileron deflection. For some aircraft projects for high speed cases an aileron preset is applied with respect to the sizing of the actuators. Using an aileron preset the LAF is switched off but still the ailerons are already deflected.

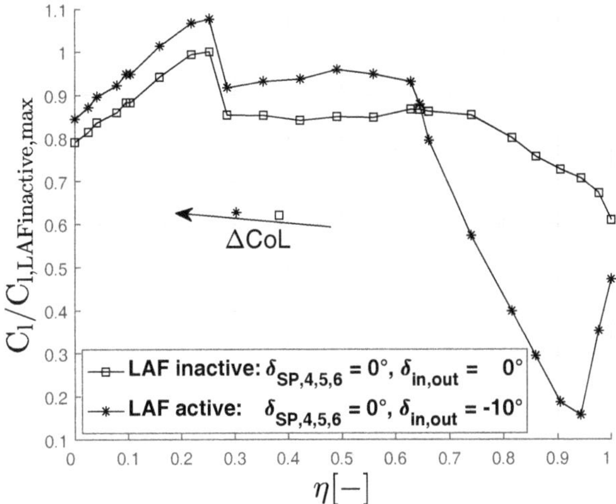

Figure 3.10: Effect of MLA parameter modification on lift distribution: 2.0g

Based on the same 2.0g example case the following Figures 3.11 and 3.12 show the impact of the same MLA parameter changes on the SMT distributions along the wing and aft fuselage section. The former shows the expected reduction of the vertical bending moment along the normalized wing semispan η and thus at its root. The latter shows the vertical shear force along the normalized x-coordinate of the aft fuselage section up to the point of fuselage and wing intersection. One can see how the elevator deflections due to the needed pitch moment compensation result in an additional force introduction. Especially the fuselage region after the normalized x-coordinate value 0.9 (location of HTP pivot) is affected where the shear force (in this example case) undergoes even a sign change. This load increase appears in particular when the CG is located more aft with respect to the fuselage payload. Hence especially cases with such a mass distribution are predestined as use cases for this kind of LAF investigations.

Based on these SMT distributions and their mentioned alterations a multi-objective optimization is one possible approach to find the best set of aileron and spoiler deflection angles. Here the objective is a vector that consists of a set of competing objectives which need to be minimized. In the patent document [Lew89] which regards a manoeuvre load alleviation system it is written that "the reduction in structural weight is provided from a reduced level of bending moment at the wing root" with the argument that "the bending movement is the principle determinator of structural strength requirements and thus weight of the aircraft wing". In this sense the wing root bending moment $M_{x,root}$ and

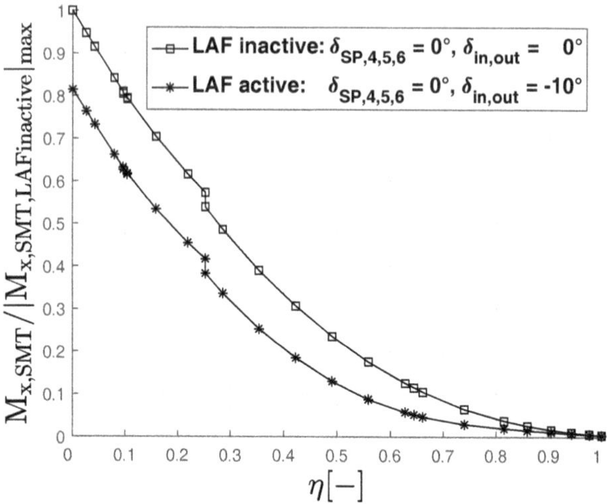

Figure 3.11: Effect of MLA parameter modification on wing SMT M_x: 2.0g

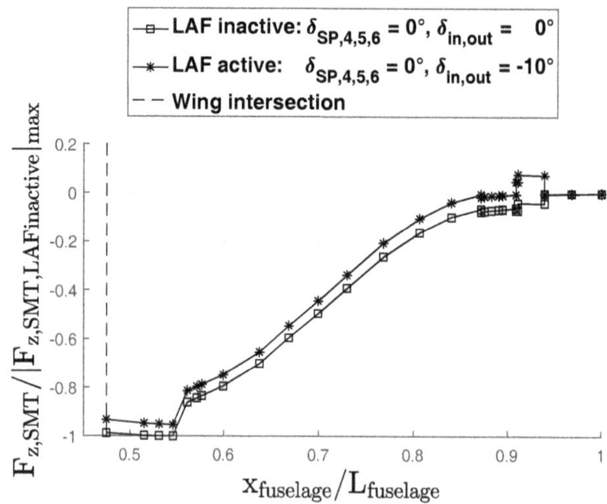

Figure 3.12: Effect of MLA parameter modification on aft fuselage SMT F_z: 2.0g

additionally the torque $M_{y,root}$, which need to be minimized, are included in the vector of objectives. In addition, the vertical shear force at the HTP pivot $F_{z,HTPpivot}$ is included and needs to be minimized, too. The quantity $F_{z,HTPpivot}$ is chosen as one of the objectives because the additional loads introduction into the fuselage is a consequence of the MLA parameter variation and the resulting pitch moment compensation. Indeed, the additional loads introduction at the HTP pivot is the main reason for the structural assessment (mass penalty estimation) of the aft fuselage region. In future research and with increasing maturity of the MDAO process also competing objectives from several disciplines of the process chain can be included into the vector of objectives. Especially the mass of components like the wing and fuselage are of interest. Currently the proposed optimization formulation consists of quantities that are available during the loads calculation process only. In this sense and using the same 2.0g example case the optimization problem can be formulated as follows:

$$\min_{X\epsilon\{\delta_{in},\delta_{out},\delta_{SP_{4,5,6}}\}} \quad [M_{x,root}(X), M_{y,root}(X), F_{z,HTPpivot}(X)]^T \tag{3.6}$$

$$\text{subject to} \begin{cases} -15^\circ \leq \delta_{in} \leq 0^\circ \\ -15^\circ \leq \delta_{out} \leq 0^\circ \\ -9^\circ \leq \delta_{SP_{4,5,6}} \leq 0^\circ \end{cases} \tag{3.7}$$

Here based on SMT values the objective is to minimize the vertical bending moment $M_{x,root}(X)$ and torque $M_{y,root}(X)$ at the wing root using different deflection angles for the LAF which are written in the vector of design variables X. In addition, as a competing objective the vertical shear force $F_{z,HTPpivot}(X)$ at the HTP pivot needs to be minimized in order to keep the load increase at the lowest possible level. The design space of the optimization variables, which consist of inner and outer aileron deflection angles δ_{in}/δ_{out} and outer spoiler deflection angles $\delta_{SP_{4,5,6}}$, is limited by upper and lower boundary values. As mentioned before technically for instance the ailerons can be deflected upwards by -25°and the spoilers by -30°when used as speedbrakes. However for the idea of optimization the range of deflection angles is chosen very closely to the recommended values as shown before in Figure 3.9. For the proposed multi-objective optimization problem with its competing objectives a Pareto optimal solution has to be found. The competing objectives here result in multiple optimal solutions which then form the Pareto frontier for one correlated load case each. Taking into account the competing objectives one gets a non-dominated solution which means an improvement at the wing root leads to a disadvantage at the aft fuselage and vice versa. The idea in building this sequential MDAO process is to quantify these effects in terms of SMT values and mass changes. Based on this quantification an improved LAF parameter setting can be proposed.
Once the multi-objective optimization problem is set-up it can be solved using the Matlab® optimization toolbox. In his book [Mes15] the author provides an overview of this optimization toolbox and its use in practice. In addition, more pieces of information on how to solve multi-objective optimization problems using evolutionary algorithms are given in the book [Deb01].
Another idea (but for future research) is to consider results of the loads calculation process over different time steps and thus to be able to perform a full EFCS optimization with all its constraints and objectives. This EFCS optimization can be performed for instance

by linear controller *gain scheduling* (cf. [LL00]) or a more advanced *multi-objective parameter synthesis* (cf. [Joo96]). *Gain scheduling* is a widespread and well-known method for optimizing a non-linear control system. Using this approach a set of linear controllers is applied whose gain values are optimized for different operating conditions (cf. [LL00]). With respect to the design of the aircraft flight control system plenty of gains have to be tuned for different conditions like pitch control or landing. In order to vary the gain values properly measurable output data, which indeed is noisy data, and also the control surface deflection angles are sent back to the controller using a feedback loop (cf. [BAL11]). The measured data contains especially the Mach number, dynamic pressure, flight altitude and calibrated airspeed. The authors of the article [LL00] provide an overview of the basic theory regarding linearisation approaches for the purpose of *gain scheduling*. In the journal paper [BAG97] the authors demonstrate the use of dynamic pressure driven *gain scheduling* by applying a linear parameter varying technique on the controller of a high performance aircraft. Also the authors of [NRW93] highlight some theoretical aspects and demonstrate *gain scheduling* using H-Infinity methods for designing an autopilot system for pitch control. H-Infinity methods are especially used for designing robust controllers by solving an optimization problem. Further in [Joo99], [Joo96] and [JBL+02] a software suite is presented for the multi-objective optimization of lateral and longitudinal controllers using multiple parameters with special attention paid on performance and robustness. This *multi-objective parameter synthesis* for controller design is able to provide Pareto optimal solutions for competing objectives. In this sense parameters like filter gains and time constants of the lateral controller are tuned using a direct optimization approach. On the other hand for the optimization of longitudinal controller a linear-quadratic-regulator synthesis method is applied in which the weighting matrices are the design variables. In [Joo99] H. D. Joos refers to these weighting matrices as synthesis parameters because in his opinion in contrast to the classical controller optimization the gains are not tuned directly. In [LJ01] H. D. Joos and G. Looye present an application in which an attitude control law for a civil aircraft is designed by a dynamic inversion approach which is combined with *multi-objective parameter synthesis* in order to improve the robustness.

4 Structural property optimization and mass penalty estimation

In this chapter the methods for the assessment of the impact of LAF parameter modifications on the structural properties are described. The results for both wing and fuselage components are driven by the modified SMT loads due to the LAF parameter modifications. For the structural property optimization of the wing covers two different approaches based on the fully stressed design algorithm are implemented and investigated with respect to possible mass reductions. The first approach is an alteration of the stress ratio based fully stressed design algorithm and uses allowable stress values which need to be defined by the user. The intention of this altered approach is to work with a better convergence behavior and to be able to run simulations for a high number of load cases and structural elements in a short time frame. The second approach is based on the idea of applying the same scaling algorithm as the one that is commonly used when working with the fully stressed design approach. However instead of relying on stress values only the idea here is to go one step further and to apply structural reserve factor values. The reserve factor values are calculated by an Airbus structural sizing software in which different failure criteria and allowable values for isotropic materials are already implemented. This second approach is more accurate but still the calculation of reserve factor values for a high number of load cases and structural elements is computationally expensive and hence time-consuming, too. The aft fuselage section is analyzed for possible increases of the structural properties due to the additional loads introduction at the HTP pivot only. Here the scope of the analysis lies in the quantification of a possible structural mass increase only. Due to the limited time frame of this PhD thesis the idea is to perform this mass penalty estimation of the aft fuselage section based on FEM internal loads. These internal loads are used to build 1- and 2-dimensional envelopes and to compare the envelopes of current load cases (with activated MLA) against those ones of baseline cases (with de-activated MLA). The exceedance of current envelopes are calculated and used as a measure for the mass penalties. Note that the used wing and fuselage models, which are provided by Airbus, are already statically sized based on a certain set of load cases and failure criteria. Here the applied load cases and failure criteria are a subset of the original ones only. Hence the idea is to make a relative quantification of the mass reductions and mass penalties with respect to a generated baseline model which is explained in the forthcoming chapter 6.

4.1 Modeling aspects of global FEM representation

One major aspect for the assessment of the impact of LAF parameter modifications on the aircraft structure is a discrete FEM representation of the wing and fuselage. In this sense the previously calculated SMT values are applied to the FEM models in order to perform a linear static analysis. This step yields different output results and in particular the

fluxes (forces per width) and stress values for the different elements and considered load cases. For this purpose the FEM models of the fuselage and the left wing are provided in MSC Nastran® compatible ASCII text files. The FEM models are parametrized based on these MSC Nastran® input files and the structural properties are then modified by self-written Matlab® code.
The FEM models, which are used here, are described in the following. Both the FEM model of the wing box and the fuselage are modeled according to the same principles. Here only Bar, Beam, Rod, Tria3 and Quad4 finite elements are used for the purpose of generating a global representation of the different aircraft components. Thus the linear static analysis is computationally inexpensive even if larger components are computed. The resulting fluxes and stresses of these finite elements are used for the purpose of strength assessment. For these so-called *models for strength assessment* a relatively fine representation is used in order to achieve a correct internal loads distribution together with an accurate representation for strength assessment. Note that for future research the resulting stress values of these global FEM representations can also be used as input for later structural fatigue life analysis and residual strength analysis of damaged structures (cf. [BB06]). The global FEM representations that are used here apply modeling aspects which are similar to the ones of those FEM models that are shown in the publications [OH17], [KDDBH19], [SKDS13] and [FWFH14]. The most significant aspects that are indeed incorporated in the used models are explained below. Note that in aeronautics industry detailed FEM models are also available but this kind of representation is used for local analysis of structural parts only and hence is not covered here.

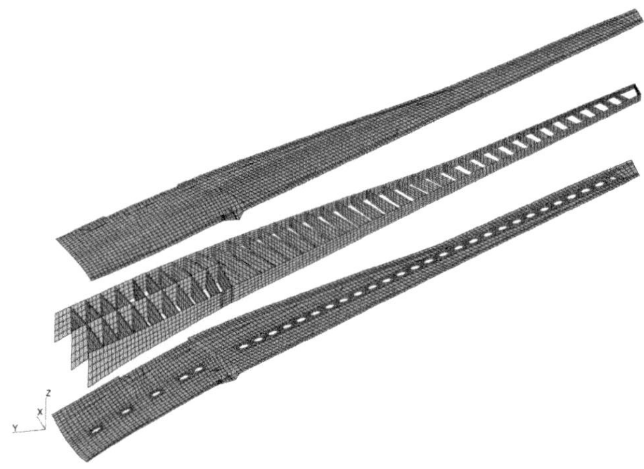

Figure 4.1: Global FEM of generic aircraft wing box model - exploded view of skin cover panels

An exploded view of the inner and outer wing box and their skin covers is shown in Figure 4.1 whereas the stringers can be seen in the appendix in Figure A.4 and part of the aft fuselage model in Figure A.5. The figures do not show the FEM models that are

used here but instead FEM models of a generic long range aircraft which are used by Airbus and partnering institutes for different research projects. As stated in [RQ14] "the finite element model is realised with shell and beam elements, and structure is meshed coarsely with relatively fine discretisation in regions where high gradient of stress" occurs. Additionally, as mentioned in [FWFH14] "in order to determine the discretization error of the analysis, it is necessary to perform a mesh convergence study". After a certain level of mesh convergence is achieved the stress values do not change significantly anymore. Regarding the wing cover discretization generally two Quad4 elements are used between two ribs and one element between two stringers. For the fuselage it is one element between each stringer and frame. This discretization is sufficient enough for a proper representation of the global stiffness and to calculate correct internal loads and displacements.
Mainly the wing skin panels need to sustain the shear stresses (cf. [Meg10], [Sun98]). Due to their low resistance against compression loads the skin panels are stiffened by additional stringers which then carry the axial loads. The contribution of the stringer and spar flange cross sections to the complete wing box cross section is small (cf. [Meg10]) and accordingly in the case of wing bending the stress changes across the stringer cross sections are assumed to be neglectable. Indeed, the stringers and spar flanges carry most of the axial stresses (cf. [Meg10], [Sun98]). The contribution of the stringers to the total wing bending stiffness is minor and hence the transverse shear loads are distributed from the stringers to the wing ribs which again distribute these loads to the wing spars (cf. [Sun98], [ACY$^+$10], [RQ14]). In the paper [ACY$^+$10] different modeling techniques for the wing cover stiffeners are highlighted. For instance the stringers can be modeled with beam elements, or a so-called *"smearing technique"* can be applied in which the stiffening effect is modeled into the shell elements.
The fuselage shell is stiffened by stringers and additionally supported by frames (cf. [Sun98]). The fuselage shell needs to sustain the shear stresses (and hoop stresses if cabin pressure is applied) and as written in the book [Sun98] "the stringers carry bending moments and axial forces" whereas the frames are used "to shorten the span of the stringers between supports in order to increase the buckling strength of the stringer".
The global FEM representation of the aircraft involves the primary structure only. For this the skin panels of the two considered components are modeled using Quad4 and Tria3 elements in which the former are predominantly used. Just a small number of the latter are used in order to mesh inhomogeneous regions of the structure which cannot be modeled properly with Quad4 elements. The use of Tria3 elements for inhomogeneous regions is just a means of keeping the discretized models as close as possible to their real geometries. Note that the Tria3 elements apply linear ansatz functions and yield less accurate results than the Quad4 elements which apply bi-linear ansatz functions (cf. [Ste12]). In this sense the fuselage floor, wing ribs and spars are also modeled using these kinds of 2-dimensional elements. The finite elements that are used for the modeling of these structural parts are assigned with shell properties which means that 3 translational and 2 rotational degrees of freedom are represented. In addition, a thickness value has to be assigned to each shell property definition which indeed is an important parameter for the structural property optimization. Note that the skin panels with manholes at the lower wing cover are excluded from the optimization. Especially the high stress concentrations at these hole regions of the lower cover require a more detailed analysis for the purpose of proper sizing. Hence with respect to the structural assessment of LAF parameter variations the analysis of skin panels with manholes is not taken into account here.

The stringers of the upper and lower wing covers are modeled with bar elements. The cross section is not tapered and further the complete six degrees of freedom are covered. Comparable to the skin panel discretization one stringer between two ribs is modeled with two concatenated bar elements. The driving parameters for the structural property optimization are the dimensions of the bar cross sectional profile and the resultant moments of inertia in vertical (I_y) and lateral (I_z) direction. Note that by default the shear coefficients for the bar properties are set to infinite which leads to zero valued transverse shear flexibility and thus the bar elements behave according to Euler-Bernoulli theory.
The fuselage frames are modeled with beam elements and here just one element is used for one frame between two stringers. In addition, one bending direction is kept in the FEM modeling only. For this purpose the second moment of inertia is set to zero. Also no torque is considered for the fuselage frames. Next, the FEM representation of the fuselage stringers is based on rod elements. With respect to this the stringer cross sectional area is the relevant parameter for the mass penalty estimation. Again just one element is used for one stringer between two frames.
The global FEM representation is a first step towards the strength assessment of the aircraft components, i.e. the wing covers. In this sense the calculated internal loads and stress values are input for the forthcoming structural sizing of dedicated components which is performed with additional Airbus internal sizing software that includes various failure criteria. The considered failure criteria are explained in the subsection 4.4.2.
The analyzed regions of the considered long range aircraft model are built of metallic material and in this sense property values of an aluminum alloy are used in the FEM model definition. Here an aluminum alloy of the class 2024 is employed whose main alloying elements are with values up to 4.9% copper and 1.8% magnesium. The applied values of these material properties are listed in the appendix in Table A.1. Also the values for the material density, yield and tensile strength are given in the same table. Here especially the density value is important for the estimation of the structural mass. The strength values are more relevant for the mass penalty estimation and the structural property optimization. In the scope of this thesis the reference values for the used aluminum alloy are not taken from Airbus internal sources but are rather based on the values that are listed in [Inc17].

4.2 Global FEM external loading

The external loading of the global FEM models of both wing and fuselage is based on Airbus methods. The methodologies for the external loading are developed by Airbus internally but then integrated into the MDAO process in the frame of this thesis. The SMT values along each of the components which are available after the loads calculation process are the basis for the external loads application. These values are applied in form of discrete forces and moments to each of the global FEM models but in two different ways. Here the forces and moments are applied to the wing component by using RBE3 elements which are placed in the plane of the ribs along the spars. On the other hand a so-called *unitary load case* approach is used for the fuselage component. This approach is based on a certain set of *unitary load cases* whose superposition has to equal the targeted SMT values along the fuselage. Both methodologies are explained in the following subsections.

4.2.1 Fuselage unitary load case approach

The core idea of the *unitary load case* approach that is used for the external loading of the fuselage global FEM model is the superposition of a certain subset of the *unitary load cases* in order to reach reference SMT values. In this sense a linear combination of previously unknown scaling factors and the predefined *unitary load cases* is performed in which each of these combinations is the equivalent of the external loading of a total load case. The predefined *unitary load cases* are based on the mass distribution of the fuselage secondary structure and are available in form of Nastran® discrete forces and moments that are applied to the grid points of the FEM model. Each *unitary load case* along the fuselage has a physical meaning and relates to real structural parts like for instance passenger seats, galleys, lavatories and hatracks. One example for a *unitary load case* is the *passenger load* which involves next to the passenger weight also the weight of the seat, entertainment system, hand luggage and life vest. Exemplary a mass of around $100kg$ per *passenger load* is assumed which is then multiplied by $1g$ and thus yields a unitary load of around $981N$ in downwards direction.
Once these mass based *unitary load cases* are defined for one aircraft, constrained by Airbus specifications and experience values, they can be adapted to another one easily. Note that their number can differ from one aircraft model to another one dependent on the intended accuracy. Indeed, the higher the level of detail is chosen, the more computational expensive is the FEM simulation.
A trivial iterative factorization process is performed in order to find the right scaling values for the linear combination. Here the computational effort can be reduced by preselecting a subset of *unitary load cases* which are relevant for the target load case. For instance *unitary load cases* that consist of a lateral force only can be excluded from the factorization process that aims to reach target values of a vertical load condition. The factorization process ends when a certain accuracy during the mean square error minimization is achieved or the maximum number of iterations is reached.

4.2.2 Wing box loading

The loading of the inner and outer wing box is based on a direct SMT approach. For this approach RBE3 elements are modeled in the plane area of the wing ribs starting from the first inner wing box up to the tip. These kind of rigid body elements are so-called interpolation elements which do not introduce additional stiffness into the structural FEM model. The previously calculated SMT loads (3 force and moment values each at dedicated cross sections) along the wing are transformed into discrete forces and moments that are applied to the RBE3 master nodes which are located along the wing reference axis. Note that the effect of high lift devices is not considered in the loads calculation process here. Accordingly also the wing global FEM model does not include high lift components and hence the loads application procedure for these structural parts is not considered here. Indeed, the loads calculation process includes the fuel loads, too. However due to the limited time frame of this PhD thesis the fuel loading on global FEM level is not implemented.

4.3 Linear static analysis and results management

The output results of the linear static analysis which is performed on the global FEM model are essential for the forthcoming structural property optimization and mass penalty estimation. The complete set of results is saved in a Nastran® binary op2 file format which is advantageous due to its low memory consumption. An op2 reader is developed at the Institute of Aircraft Design and Lightweight Structures of the Technische Universität Braunschweig based on the Python programming language and integrated as a core functionality into the MDAO process. Generally speaking the op2 reader converts different blocks of the op2 file into a readable hdf5 file format. Here especially the element forces and stresses as well as grid point forces are relevant. This file format is especially suited for huge data which can then be stored in form of multiple dimensions. Another aspect is the use of Matlab® as the main programming language in this prototyping phase (see also chapter 2.3) across the MDAO process chain. Due to aspects of tool harmonization and the inbuilt functionality of Matlab® to read the hdf5 file format this way of results management is chosen. The flowchart in Figure 4.2 shows again the implemented steps from the linear static analysis to the Matlab® variable that contains the output results and can be processed further inside the MDAO process.

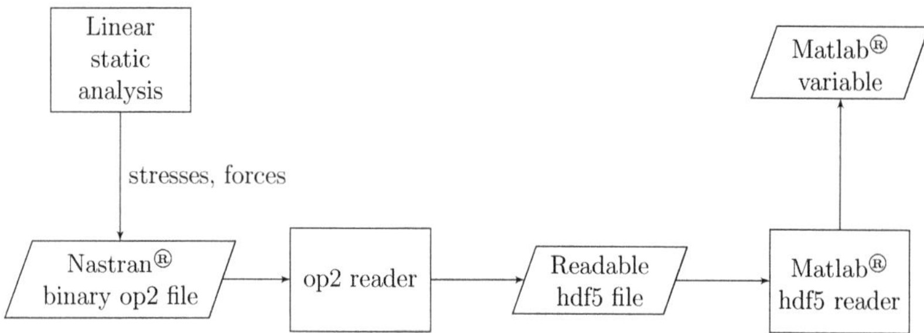

Figure 4.2: Linear static analysis output results management

Different sets of results and result files are necessary for the forthcoming structural property optimization and mass penalty estimation. As mentioned before the first approach that is used for the structural property optimization of the stiffened wing covers applies an alteration of the stress ratio based fully stressed design algorithm. This approach requires the element stress values which are retrieved from the generated Matlab® variable. The second approach that is used for the structural property optimization of the stiffened wing covers applies a combination of both the commonly used fully stressed design scaling algorithm and the structural reserve factor values. This approach requires the element forces and grid point forces which are imported to the Airbus internal sizing software by using the generated op2 file directly. Finally the mass penalty estimation process of the aft fuselage region requires the element forces and stress values which are retrieved from the generated Matlab® variable, too.

4.4 Wing structural property optimization

The purpose of the structural property optimization of the stiffened wing covers is to assess the possible mass reduction due to the previously performed LAF parameter modifications. Two different fully stressed design approaches are implemented and tested. The first one is an alteration of the stress ratio based fully stressed design approach and the second one includes structural reserve factors for different failure criteria in the fully stressed design scaling algorithm. Both approaches are explained in more detail in the following subsections. The two approaches have in common that they use the same vector of design variables $\mathbf{x}_i$ which are indeed constrained with lower and upper boundary values that are given in $\mathbf{x}_l$ and $\mathbf{x}_u$ accordingly (cf. equation 4.1). Here the vector of design variables $\mathbf{x}_i$ contains the wing skin panel thickness values and geometric parameters of the stringer cross sections.

$$\mathbf{x}_l \leq \mathbf{x}_i \leq \mathbf{x}_u \tag{4.1}$$

In addition, each of these iterative processes ends when a certain mass convergence criterion is reached which is shown in equation 4.2. Note that re-sizing is performed once per total iteration loop and the mass convergence is checked in the forthcoming feedback loop module. Here the change of total mass of the wing component from one iteration to the next one is used. Note that the mass value is calculated with the MSC Patran® Grid Point Weight Generator which is explained in more detail in chapter 5.

$$\triangle m_{tot} \leq \varepsilon_m \tag{4.2}$$

As stated in [Joh02] commercial software like MSC Nastran® offers a fully stressed design solution, however the scope of its application is limited to the shell thickness and the cross section area of rod elements. In the scope of this thesis both the wing cover skin thickness values and the geometric parameters of the stringer cross sections, which affect the stringer moments of inertia accordingly, need to be analyzed and modified. Hence an own fully stressed design implementation is programmed. In this sense the available global FEM model of the upper and lower wing covers is completely parametrized and thus each shell and bar element has its own property entry using MSC Nastran® property cards. Note that the implemented process provides the possibility to define optimization regions by providing a list of dedicated finite elements. Thus dedicated regions of the wing covers can be optimized locally, too.

4.4.1 First approach - Fully Stressed Design using alteration of stress ratio based algorithm

The MDAO process is built sequentially and hence the loads calculation process with its EFCS *parameter* variation and the here described structural property optimization are performed in two separated steps. The optimization of the structural properties is purely done inside its own dedicated module but indeed uses the values from the loads calculation module. In contrast to gradient based optimization algorithms which rely on derivatives and minimize or maximize an objective function the fully stressed design approach is based on stress ratio values only. As shown in equation 4.3 in general each element of the vector of updated design variables $x_{i,new}$ is the result of the product of its old value and the ratio of its related response value y_i and a corresponding allowable y_{allow}. The value of the exponent α steers the convergence of this algorithm and can take values between 0.0 and 1.0 (cf. [Joh02]). Higher values in the exponent lead to faster convergence. Indeed, this general formulation is adapted for the use of stress ratios. As stated in [Raz65] and taking the skin thickness as an example the algorithm increases the value when the current resulting stress is higher than the allowable and decreases the thickness value for the opposite case.

$$x_{i,new} = x_{i,old} \left(\frac{y_i}{y_{allow}} \right)^{\alpha} \tag{4.3}$$

Even for a large number of design variables the fully stressed design approach is still computationally inexpensive and less time consuming than a mathematical optimization. As mentioned in [Joh02] these two major advantages make this stress ratio algorithm well suited for a quick initial sizing of the structure (5 to 10 iterations recommended) before an expensive and time consuming mathematical optimization is applied. Note that no mathematical proof exists for the fully stressed design algorithm which shows that its application leads to minimized mass values. However in [Raz65] and [Joh02] it is highlighted that each design parameter reaches its allowable stress value for at least one of the total set of applied load conditions. On the other hand in the same article [Raz65] it is mentioned that an optimum is not reached necessarily in the case of indeterminate structures. Further in [Fuc16] it is demonstrated on a basic truss model example that a fully stressed state is not achievable for redundant structures.
The implemented fully stressed design approach for the wing cover skin panels is based on the work of H. Miura who presents in his article [Miu90] an alteration of the fully stressed design algorithm with a better convergence behavior. The main idea that is presented in [Miu90] is a modification of the stress ratio algorithm for the case of plate and shell structures, i.e. for the coupled occurrence of in-plane membrane and out-of-plane bending loads. As stated in [Miu90] the idea of this altered fully stressed design approach is to improve the "oscillatory iteration histories" when optimizing the skin panels and both in-plane loads as well as "transverse bending loads" are taken into account. The commonly used fully stressed design stress ratio algorithm as shown in equation 4.3 shows a good convergence behavior with an α value of 1.0 when applying pure in-plane loads and with an α value of 0.5 when applying pure bending loads (cf. [Miu90]). Hence the idea of this altered fully stressed design approach is to avoid a predefined exponent value α and instead steer the convergence behavior during the analysis based on current stress values. For this approach the von Mises criterion (cf. equation 4.4) is used and the calculated values at the top and bottom side of each shell element are compared with an allowable value. Each part of the von Mises equation is substituted by its explicit form which involves stress values due to both the in-plane loads and the bending loads.

$$\sigma_{vonMises} = \sqrt{\sigma_x^2 + \sigma_y^2 - \sigma_x \sigma_y + 3\tau_{xy}^2} \tag{4.4}$$

For instance equation 4.5 shows the relation between the normal stress $\sigma_{x,top}$ at the top side of the shell element and the related axial force N_x and bending moment M_x. Here the variable for the thickness of the shell element is t. The relations for $\sigma_{x,bottom}$, $\sigma_{y,top}$, $\sigma_{y,bottom}$, $\tau_{xy,top}$ and $\tau_{xy,bottom}$ are built according to the same principle.

$$\sigma_{x,top} = \frac{N_x}{t} + \frac{6M_x}{t^2} \tag{4.5}$$

After several transformations the core equation that describes the update of the skin thickness value is derived (cf. equation 4.6). Here the stress ratio is replaced by the scaling factor α^* which is the solution of the algebraic equation 4.7.

$$t_{new} = \alpha^* \, t_{old} \tag{4.6}$$

$$A{\alpha^*}^4 - B{\alpha^*}^2 \pm C\alpha^* - D = 0 \tag{4.7}$$

The content of the coefficients A, B, C and D is listed in the equations 4.8 to 4.14. These coefficients take into account the allowable stress value and normal as well as shear stress values of the top and bottom side of the shell elements.

$$A = \sigma_{allow}^2 \tag{4.8}$$

$$B = \sigma_{Sx}^2 + \sigma_{Sy}^2 - \sigma_{Sx}\sigma_{Sy} + 3\tau_{Sxy}^2 \tag{4.9}$$

$$C = 2\sigma_{Sx}\sigma_{Dx} + 2\sigma_{Sy}\sigma_{Dy} - \sigma_{Sx}\sigma_{Dy} - \sigma_{Sy}\sigma_{Dx} + 6\tau_{Sxy}\tau_{Dxy} \tag{4.10}$$

$$D = \sigma_{Dx}^2 + \sigma_{Dy}^2 - \sigma_{Dx}\sigma_{Dy} + 3\tau_{Dxy}^2 \tag{4.11}$$

The skin panels are modeled with at least two Quad4 elements each in order to represent an appropriate bending behavior. Thus the altered fully stressed design algorithm is properly applicable with the used global FEM representation of the wing covers.

$$\sigma_{Sx} = \frac{(\sigma_{x,top} + \sigma_{x,bottom})}{2}, \; \sigma_{Dx} = \frac{(\sigma_{x,top} - \sigma_{x,bottom})}{2} \tag{4.12}$$

$$\sigma_{Sy} = \frac{(\sigma_{y,top} + \sigma_{y,bottom})}{2}, \; \sigma_{Dy} = \frac{(\sigma_{y,top} - \sigma_{y,bottom})}{2} \tag{4.13}$$

$$\tau_{Sxy} = \frac{(\tau_{xy,top} + \tau_{xy,bottom})}{2}, \; \tau_{Dxy} = \frac{(\tau_{xy,top} - \tau_{xy,bottom})}{2} \tag{4.14}$$

The implemented algorithm for the wing stringers is based on the usual stress ratio algorithm. Here the maximum stress value of both grid points of each bar element is chosen and compared with the allowable value. The wing stringers are modeled with a J-profile cross section and their modified dimensions are shown in Figure 4.3. Here L_1, L_2

and L_3 are the height of the web, the width of the free flange and the width of the attached feet whereas L_4, L_5 and L_6 are the thickness values of each section. Accordingly the implemented fully stressed design algorithms are shown in equations 4.15 and 4.16 in which the convergence values for α are set as proposed in [Öst06] and are based on the authors simulation studies for convergence assessment.

$$L_{1_{new},\ldots,3_{new}} = L_{1_{old},\ldots,3_{old}} \left(\frac{\sigma_{x,max}}{\sigma_{allow}} \right)^{0.25} \tag{4.15}$$

$$L_{4_{new},\ldots,6_{new}} = L_{4_{old},\ldots,6_{old}} \left(\frac{\sigma_{x,max}}{\sigma_{allow}} \right)^{0.5} \tag{4.16}$$

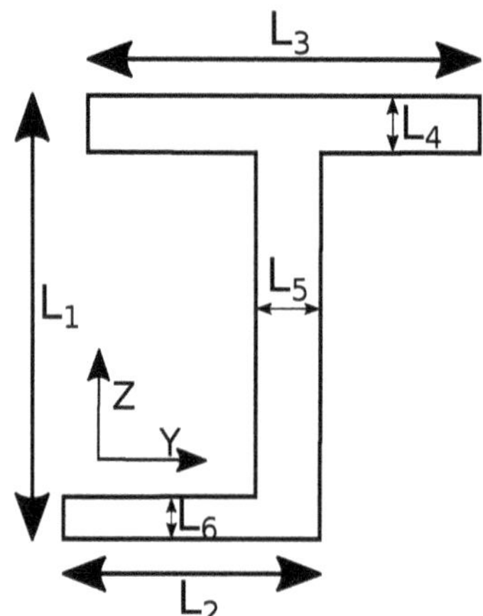

Figure 4.3: Wing J-profile stringer cross section - modified dimensions

Since the dimensions of the stringer cross section profile are modified the moments of inertia need to be recalculated, too (cf. equations 4.17 and 4.18). Both moments of inertia I_y and I_z for vertical as well as lateral bending are considered. Note that the cross section area A_i is updated after each fully stressed design iteration. The very first reference values for the dimensions of the stringer cross section and inertia values are retrieved from the model that is used inside the Airbus internal structural sizing software.

$$I_y = \sum_{i=1}^{n} I_{y,i} + \sum_{i=1}^{n} z_i^2 \, A_i \tag{4.17}$$

$$I_z = \sum_{i=1}^{n} I_{z,i} + \sum_{i=1}^{n} y_i^2 \, A_i \tag{4.18}$$

The global FEM *model for strength assessment* of the wing component is used for the linear static analysis and hence modeled with a fine discretization. According to this the

Design variables for wing box upper/lower cover	
Finite element	**Number of elements**
Quad4	2217
Tria3	90
Bar	2085

Table 4.1: Total number of possible design variables - wing box covers fully stressed design

total number of possible design variables is high. These are made up of the skin panel Quad4 and Tria3 elements as well as of the stringer bar elements. The total number of possible design variables is listed in Table 4.1. Note that here the idea is a fully stressed design optimization of already sized stiffened wing box covers due to LAF parameter modifications with the objective of a possible mass reduction. As mentioned before the here used failure criteria and load conditions are a subset of the original ones only. In this sense especially changes in the skin panel thickness can lead to a high mass reduction and hence need to be interpreted carefully. Note that those areas of the wing covers where the landing gear attachment loads are introduced are primarily sized by ground loads. Since ground loads are not taken into account here these areas of the wing covers are excluded from the fully stressed design optimization. Indeed, this kind of exclusion reduces the number of design variables.

Further in order to reduce the computation time due to the high number of design variables the fully stressed design approach for both skin panels and stringers is parallelized. The parallelization is realized with the Matlab® parallel computing toolbox which distributes the calculation to 8 different processors (time reduction by factor 8) on the Airbus cluster.
The upper and lower boundaries of the skin panel thickness and stringer cross section dimensions are based on the initial minimum and maximum values of the Airbus internal model that is used for sizing analysis. Further the allowable stress values which are applied for this approach are based on the aluminum material values as listed in Table A.1.

4.4.2 Second approach - Fully Stressed Design using structural reserve factors

The second implemented fully stressed design approach is based on the commonly used stress ratio algorithm (cf. equation 4.3) but involves structural reserve factor values for different failure criteria. In terms of computational implementation the main principles stay the same and the global FEM model is still the central focus of study. The core idea here is to replace the stress values in the scaling algorithm with reserve factor values in order to consider more failure criteria than just the von Mises yield criterion. As mentioned before the here used load conditions and sizing criteria are a subset of the original ones that are used for the sizing of the wing box. The incomplete set of load conditions and sizing criteria yield too optimistic mass reductions when applying the structural property optimization. In this sense a more conservative assessment of the possible mass reductions is performed by including the structural reserve factor values for different failure criteria. Figure 4.4 provides an overview of this reserve factor based fully stressed design approach. The process starts with the definition of the global wing FEM *model for strength assess-*

ment which is used for the linear static analysis. The global FEM properties, geometry data and results are then used as input data for the Airbus internal sizing software. For the calculation of the considered failure criteria the element forces (Quad4, Tria3, bar) and grid point forces are needed. The calculated reserve factor values are then processed in the implemented fully stressed design algorithm which updates the global FEM properties and transfers the new values to the model definition.

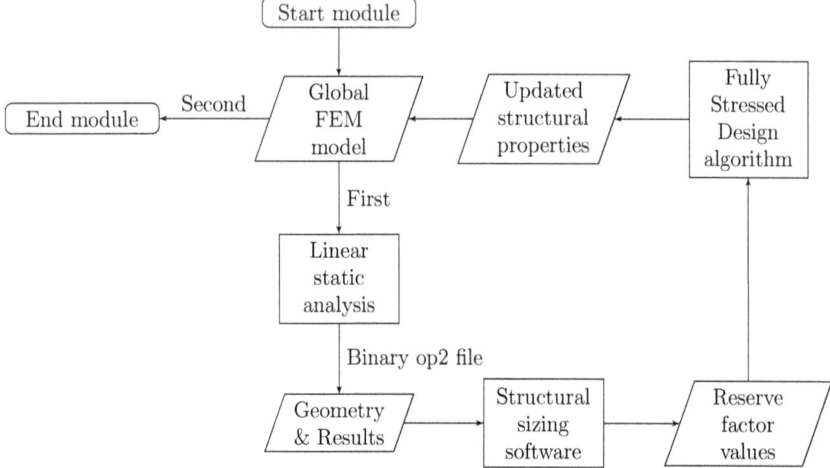

Figure 4.4: Flow chart with process overview of reserve factor based fully stressed design approach

As shown in equation 4.19 the reserve factor is generally defined as the ratio of an allowable stress value σ_{allow} and an applied stress value $\sigma_{applied}$. The reserve factor can take values higher than 1.0 (structure sustains the applied loads), less than 1.0 (structure fails), and equal to 1.0 (structure is at its limit). Each time the structural properties are modified both the linear static analysis and the reserve factor calculation need to be performed again.

$$RF = \frac{\sigma_{allow}}{\sigma_{applied}} \gtreqless 1 \tag{4.19}$$

For each failure criterion a reserve factor is calculated and for the purpose of structural property optimization using the fully stressed design algorithm the minimum value is chosen. In this sense the reserve factor values are calculated for each stiffened panel and the minimum value is used to scale up or down the dimensions of its structural components. Since the reserve factor ratio is the inverse of the stress ratio algorithm the applied equation for the skin panel thickness is implemented as shown in equation 4.20. Here again the exponent value α steers the convergence behavior. For the scaling of the skin panel thickness the α exponent is chosen by default as proposed in [Joh02] for the purpose of better convergence. Note that the MDAO interface offers the possibility to modify this convergence value. The algorithm for the scaling of the stringer cross sectional

Property	α value
t	0.9
$L_{1,\ldots,3}$	0.25
$L_{4,\ldots,6}$	0.5

Table 4.2: Default exponent values for scaling of skin panel thickness and stringer cross sectional dimensions

dimensions is implemented as shown in equation 4.21. The different values for the exponent α are again set as proposed in [Öst06]. The default exponent values for the property changes of the skin thickness and stringer cross sectional dimensions are listed in Table 4.2.

$$t_{new} = t_{old} \left(\frac{1}{RF_{min}} \right)^{\alpha} \tag{4.20}$$

The minimum reserve factor value of the considered failure criteria steers the scaling of the skin panel and stringer dimensions. However the exponent value α is not changed for the different failure criteria. The use of a different convergence rate for different failure criteria can improve the results of this fully stressed design approach. However due to the limited time frame of this PhD thesis this investigation is kept for future studies.

$$L_{1_{new},\ldots,6_{new}} = L_{1_{old},\ldots,6_{old}} \left(\frac{1}{RF_{min}} \right)^{\alpha} \tag{4.21}$$

The sizing software uses a stiffened panel idealization for the calculation of reserve factor values. The reserve factor values are calculated for different failure criteria in which the minimum result value is chosen as representative for the whole stiffened panel. Here eight failure criteria, which can be sorted into five different categories, are considered by the static strength analysis software for metallic stiffened panels. The upper cover is predominantly loaded with compression and the lower cover is mostly loaded with tension. According to this opposing way of loading the dominant failure criteria are also different at the upper and lower covers. The considered failure criteria are explained as follows.
The first category considers the strength analysis. Here the **in-plane strength** failure criterion is based on the von Mises yield stress (cf. equation 4.4) which is calculated by using the shear, longitudinal and transverse stresses. The resulting von Mises stress is then compared with the allowable material value whereupon structural failure occurs if the ultimate tensile stress is exceeded. In addition, the **net section strength** criterion is predominantly applied for the sizing of the lower wing cover under tension. Here the holed plates which form a joint need to withstand the applied ultimate loads. Holes in a plate disturb the force flow and thus the stress distribution which then leads to stress concentrations around the holes (cf. [SGN10]). The net section strength allowable value is calculated as shown in equation 4.22. Here the ultimate tensile force of the net section $F_{tu,net}$ is calculated using the ultimate tensile force F_{tu}, the ratio of net and gross section area as well as the net tension efficiency factor k_t, which can be retrieved from dedicated diagrams according to the used material (cf. [Niu99]).

$$F_{tu,net} = k_t\, F_{tu} \left(\frac{A_{net}}{A_{gross}}\right),\ with\, 0 < k_t \leq 1 \tag{4.22}$$

The effective net section area A_{net} is calculated as shown in equation 4.23. Here the part of the plate cross section area which is holed (multiplied by the number of rivet lines n_{rivet}) needs to be subtracted from the gross section area. The subtracted area is calculated using the thickness t of the plate and the diameter D of the hole.

$$A_{net} = A_{gross} - (n_{rivet} * t * D) \tag{4.23}$$

The second category considers the stability analysis of the stiffened panels due to *incomplete diagonal tension* and includes **shear buckling** of the skin, **forced crippling** of the stringer and **column buckling**. The theory of *incomplete diagonal tension* (cf. [KPL52] and [Niu99]) combines both the shear resistant and pure diagonal tension web theory. In their technical report [KPL52] the authors describe the shear resistant web as "sufficiently thick to resist buckling up to the failing load" and the diagonal tension web as "so thin that it buckles into diagonal folds at a load well below the design load". In practice most of the webs are stressed between shear and pure diagonal tension which makes it more important to apply the *incomplete diagonal tension* theory in order to calculate the stress values more accurately (cf. [KPL52]). In this sense the shear buckling failure criterion is used. The allowable value is calculated according to the approach that is explained in [KPL52] and is shown in equation 4.24.

$$\tau_{PS,allow} = \tau^*_{allow}\,(0.65 + \Delta) \tag{4.24}$$

Here the basic allowable value τ^*_{allow} can be retrieved from a dedicated diagram (cf. [KPL52]) based on the used material. The correction value Δ for the allowable shear stress is an empirical value which takes into account the "stiffening ratios involving the areas as parameters" ([KPL52]). In addition, forced crippling is also taken into account. This failure criterion is a form of local instability "on uprights [...] of open section" ([KPL52]) (i.e. stringer web or flange) that yields stress concentrations and thus crippling failure. Here the allowable value $\sigma_{FC,allow}$ is calculated as shown in equation 4.25 and is based on the work that is written in [KPL52]. Here C^* is a material dependent factor that takes the stringer geometrical shape (one- or double-sided) into account, too. As stated in [Niu99] the diagonal tension factor k can take values from 0 to 1 where $k = 0$ means a shear resistant web, $k = 1$ means a pure diagonal tension web and $0 < k < 1$ yields incomplete diagonal tension.

$$\sigma_{FC,allow} = C^* \sqrt[3]{k^2 \left(\frac{t_{flange}}{t_{web}}\right)} \tag{4.25}$$

Also column buckling is used as a failure criterion (global buckling mode) and is based on Euler's principles. In the case of column buckling ultimate loads are used and the allowable shear stress considers the compression stress due to incomplete diagonal tension.

The third category considers the **local buckling** of the skin panel. Limit loads are needed for this analysis. Here the panel can "buckle into either the local mode [...] or the wrinkling mode" ([SP56]). This *panel wrinkling* is an initial buckling mode and is regarded as a short wavelength buckling mode in the case of riveted stiffened panels and usually appears for short panels (cf. [SP56]). *Panel wrinkling* can lead to stringer crippling and thus result in failure of the stiffened panel. In this case ultimate loads are needed.
The fourth category regards the **flexural wrinkling** failure criterion which is a combination of flexural instability (column buckling) and *panel wrinkling*. Ultimate loads are needed for this criterion.
The fifth category regards column failure in which **flexural-torsional column buckling** under pure compression is used as the failure criterion (cf. [Arg54]). Ultimate loads are needed for this criterion, too.
The sizing of the wing upper cover under compression is driven by the flexural wrinkling, flexural-torsional column buckling and shear buckling critera. The sizing of the lower cover under tension is driven by the net section strength criterion. Indeed, regarding the interpretation of the sizing results engineering judgment is indispensable. Here for instance the re-sizing of the stiffened wing covers should not significantly affect the proportion of skin or stringer dominance. Note that for the lower cover fatigue and damage tolerance criteria are more relevant than the ones which are included in the static strength analysis. However fatigue and damage tolerance criteria are not handled here and hence their missing effect is unfortunately not assessed. It is also worth mentioning that the used sizing software is based on the global FEM model representation and thus its results are normally used for the assessment of sizing relevant load cases when the structural capability reaches its limits. Further in-depth strength analysis needs to be performed with detailed FEM models of the structural components.
In total the strength analysis for the upper and lower covers of the inner and outer wing box includes 888 stiffened panels. Using the Airbus internal sizing software this module is the most time consuming part of the whole MDAO process. Of course the calculation of the reserve factor values for the complete set of stiffened panels is again parallelized which lessens the computation time.

4.4.3 Fully Stressed Design - proposals for improvement

Different steps can be taken in future studies to improve the performance and results of the fully stressed design algorithm. Here just a few proposals are presented which can improve the implemented structural property optimization module of the MDAO process. The proposed ideas are not new but already widely used in the field of structural analysis. Due to the limited time frame the implementation is kept for future studies.
The first proposal is to apply a smoothing algorithm to the updated properties in order to avoid a so-called checkerboard pattern across the wing covers. Smoothing algorithms are a standard procedure when working with fully stressed design algorithms but are just not implemented here due to the limited time frame. With respect to fully stressed designs checkerboard patterns evolve from an increase of property values at highly stressed regions and vice versa. Such increase for instance of the thickness value results in higher stiffness which then itself is a reason for a stress increase. Accordingly a smoothing algorithm is presented in the article [LSX01] which is applied in the frame of topology optimization and works "in terms of the surrounding element's reference factors". This way of smoothing that is based on a weighted average is also used in both works [Öst06] and [Rie13].

Another proposal is to work with predefined sizing regions across the wing covers which also prevent checkerboard patterns and support the possible consideration of manufacturing constraints. In the journal paper [GKB09] the authors present a multi-step optimization process at Airbus in which for instance the results of the strength analysis for one structural component are applied to components of the surrounding area for the purpose of design variable reduction.

4.5 Fuselage mass penalty estimation

The activation of the load alleviation system leads to a change of the aileron and spoiler deflection angles which then yields a change of the pitch moment. This additional pitch moment needs to be compensated with elevator deflections by which an additional force is introduced at the HTP pivot. The related force introduction into the fuselage is the main reason for the mass penalty estimation of the aft fuselage region.
As shown in the last section, the structural property optimization of the wing box covers is based on different failure criteria and related allowable values which then lead to the calculation of reserve factor values. The aft fuselage section is analyzed for possible increases of the structural properties and thus of the structural mass only. Due to the limited time frame the mass penalty estimation of the aft fuselage shell (skin panels, stringers, frames) is based on internal loads exceedance values only. The global FEM internal loads are used to build 1- and 2-dimensional envelopes whereupon the envelopes that are based on the current load case set (with activated LAF) are compared with those ones that are based on the baseline load case set (with de-activated LAF). The exceedance values of the current envelopes are calculated and used as a measure for the mass penalties. In this sense the total mass penalty is the summation of the necessary mass increases of the aft fuselage skin panels, stringers and frames. Note that the fuselage mass penalty estimation is based on global FEM internal loads and is implemented as a first approach. In future studies more accurate approaches can be implemented. Especially fatigue and damage tolerance criteria, which are highly non-linear, are the most dominant sizing criteria for this region. The process flow of the implemented mass penalty estimation approach is shown in the flowchart in Figure 4.5 and is explained as follows.
The first step is the linear static analysis of the global FEM model which provides for each finite element the resulting stresses, forces and moments that are then processed with the previously described op2 reader. Afterwards the extremal stress values of each element are compared with a predefined stress threshold value that is based on the aluminum alloy material values (cf. appendix, Table A.1). The idea is to perform a first check of the stress level of each element in order to avoid misleading interpretations of the internal loads exceedance values that are calculated in the upcoming step. Indeed, some elements can have a low and thus irrelevant stress level but still yield numerically high internal loads exceedance values. In this sense a dedicated stress value (note: applied loads are limit loads) is chosen for each element type of the considered fuselage shell and is compared with the yield strength:

- Skin panel: $\sigma_{vonMises} \geq R_{p0.2}$
- Stringer: $\sigma_{axial,rod} \geq R_{p0.2}$
- Frame: $\sigma_{axial,bar} \geq R_{p0.2}$

The axial stress $\sigma_{axial,bar}$ of the frame (bar element) is the average axial stress value due to the average axial force. In the used global FEM model the location of the four stress recovery points at the bar cross section is not given in the property definition. Hence the average axial stress is used here. For the stringer (rod element) the axial stress $\sigma_{axial,rod}$ due to the axial force is used.
If the stress level of the complete set of finite elements is too low, i.e. the number of excluded elements $FE_{excluded}$ equals the number of analyzed ones $FE_{analyzed}$, then the mass penalty estimation approach is suspended because there is no need for the calculation of necessary mass increases. After this pre-selection step only the affected finite elements are considered for further analysis in order to be processed by the internal loads exceedance calculation algorithm. The exceedance values are calculated separately for each of the element types of the fuselage model. The internal loads exceedance calculation for each element type, which is based on calculated internal forces and moments, is explained as follows.

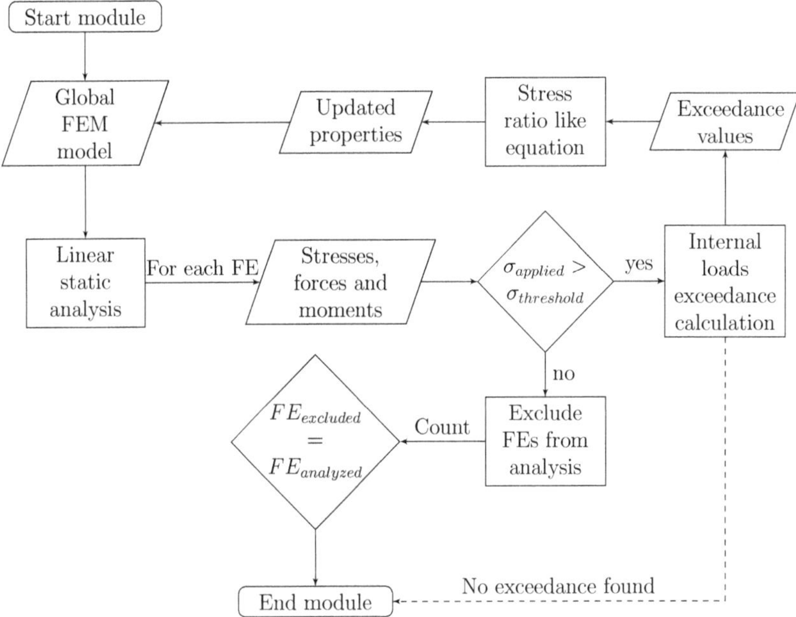

Figure 4.5: Flow chart with general overview of mass penalty estimation approach

As described in section 3.1 the considered load case set consists of manoeuvre and continuous turbulence cases. Hence the linear static analysis provides for each finite element resulting internal forces and moments for the complete load case set. In this sense the idea is to calculate the internal loads exceedance values based on extremal values, i.e. minimum/maximum internal forces and moments overall load cases, and thus perform the assessment with envelopes only. This approach also leads to less computation time which indeed increases with the number of considered load cases if the internal loads exceedance values are calculated for each single load case.

For the stringer internal loads exceedance value the extremal axial force $|N|_{max}$ of the current load case set (with activated LAF) is compared with the extremal value of the baseline load case set (with de-activated LAF). In this sense 1-dimensional envelopes are built for each stringer of the analyzed aft fuselage region and the internal loads exceedance value $\xi_{Stringer}$ is calculated as the ratio of the current and baseline envelope value (cf. equation 4.26).

$$\xi_{Stringer} = \frac{|N|_{max,current}}{|N|_{max,baseline}} \tag{4.26}$$

As explained before the properties of the wing covers are modified based on the fully stressed design stress ratio algorithm (cf. equation 4.3). Here the fuselage stringer cross section is scaled up according to the same principle if $\xi_{Stringer} > 1.0$:

$$A_{new} = A_{old}\left(\xi_{Stringer}\right)^{\alpha} \tag{4.27}$$

The exponent value α is used again to steer the convergence behavior. Here the same values are applied as for the property scaling of the wing covers (cf. Table 4.2).
The internal loads exceedance value of the frame is assessed by calculating the extremal value of the average bending moment based on the bending moment value M_y at both grid points of the bar element. As shown in equation 4.28 the extremal average bending moment value of the current load case set is then compared with the corresponding value of the baseline load case set.

$$\xi_{Frame} = \frac{|0.5\sum_{i=1}^{2} M_{y,Pi}|_{max,current}}{|0.5\sum_{i=1}^{2} M_{y,Pi}|_{max,baseline}} \tag{4.28}$$

Figure 4.6 shows schematically the cross section of a fuselage Z-profile frame as it is used here. The design and assembly of the cabin as well as manufacturing constraints restrict the possibilities to modify the Z-profile frame dimensions. The frame is connected to a shear clip which is connected to the skin. Hence the only proper modification here is limited to the thickness value L_5 of the frame web.

Accordingly the thickness of the frame web is updated (if $\xi_{Frame} > 1.0$) as shown in equation 4.29 and the convergence value α is again chosen in the same manner as for the wing stringer based on the studies which are shown in [Öst06].

$$L_{5,new} = L_{5,old}\left(\xi_{Frame}\right)^{\alpha} \tag{4.29}$$

As mentioned earlier the FEM model of the fuselage frames considers one bending direction only and according to this the moment of inertia for vertical bending has to be calculated only. The calculation of the initial moment of inertia and the updated one is implemented into the software code of the mass penalty estimation approach.
Finally the internal loads exceedance value for the skin panel is based on a 2-dimensional convex hull which is based on the in-plane forces of the Quad4/Tria3 element. Three different envelopes are built for each element, which are based on the possible combinations of the in-plane forces N_x (circumferential), N_y (axial) and N_{xy} (shear). Note that although

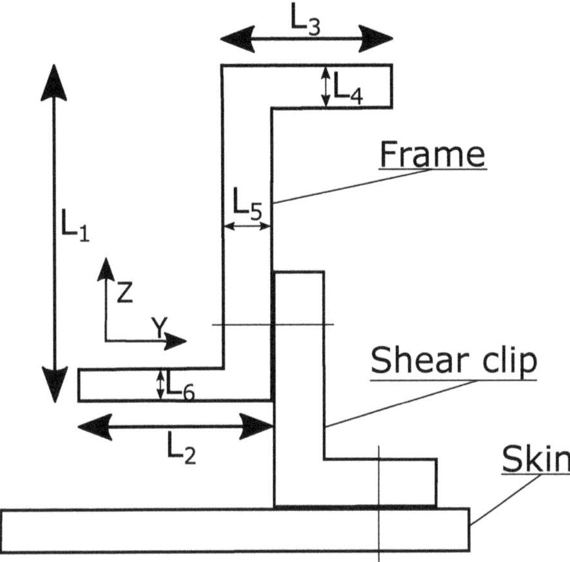

Figure 4.6: Fuselage Z-profile frame cross section - potential dimensions for modification

the side shells are predominantly affected by shear loads and the upper/lower shells by axial loads still the three possible combinations are used for the complete aft fuselage skin panels. The three possible combinations of in-plane forces are as follows:

- N_y & N_{xy}
- N_x & N_{xy}
- N_x & N_y

In this sense one convex hull for one Quad4/Tria3 element is built using the cloud of points which represent the 2-tuples of internal forces. The number of 2-tuples equals the number of calculated load cases. Figure 4.7 shows an example for a 2-dimensional convex hull based on the in-plane forces N_x and N_{xy}. For this figure the values of the forces along the abscissa and ordinate are normalized each to the absolute maximum value of the baseline load case set. As can be seen in the figure one envelope is built for each load case set (with active and inactive LAF). In order to calculate the internal loads exceedance value for the skin panel ξ_{Skin} polar coordinates are used. A polar coordinate system is drawn into Figure 4.7 in order to better illustrate this procedure. First the 2-tuple with the longest distance from the origin is determined by expressing this value as a radius including the quadrant of its location. This step is performed for both load case sets. The value of ξ_{Skin} is then calculated as the ratio of the radius of the current load case set $r_{2,current}$ and the radius of the baseline load case set $r_{1,baseline}$ (cf. equation 4.30).

$$\xi_{Skin} = \frac{r_{2,current}}{r_{1,baseline}} \tag{4.30}$$

If an internal loads exceedance appears, i.e. $\xi_{Skin} > 1.0$, then the thickness value of the skin panel is updated according to equation 4.31, in which the exponent value α is again based on the recommendation that is given in [Öst06]. As can be deduced from equation 4.31 the updated skin panel thickness value t_{new} is always higher than its previous one t_{old} since the exceedance value ξ_{Skin} is larger than one. In this sense the intended objective is achieved and the fuselage properties are scaled up and mass penalties are estimated only.

$$t_{new} = t_{old} \left(\xi_{Skin}\right)^{\alpha} \tag{4.31}$$

In the end the global FEM model is updated with the new property values and the steps from the linear static analysis onwards are rerun until no further internal loads exceedance occurs. The mass value itself, that is the basis for the resulting penalty, is calculated with the MSC Patran® Grid Point Weight Generator which is explained in the next chapter. At this point it is mentioned that next to the finite element geometry and properties the density value is another driving parameter for this weight generator. Here again the density value of the introduced aluminum material is used. Since mass penalties are estimated only no lower boundaries for the modified dimensions are set and further no upper boundaries are set in order to assess the extend of this proposed approach.

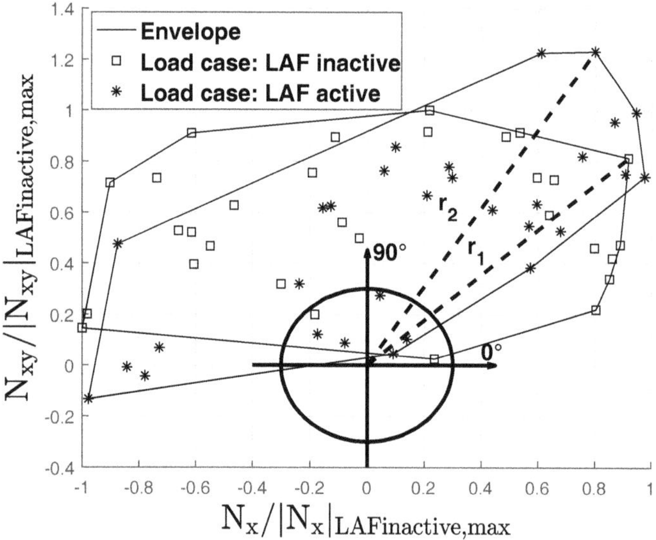

Figure 4.7: Internal loads exceedance calculation: 2D convex hull - example envelope for a Quad4 element using normalized internal forces for 30 load cases

In contrast to the structural property optimization of the wing the mass penalty is estimated on the basis of the default global FEM model of the fuselage only. This means that no updated fuselage properties are sent back to the loads calculation model via

Design variables for aft fuselage mass penalty estimation	
Finite element	**Number of elements**
Quad4	4204
Tria3	247
Bar	4313
Rod	3565

Table 4.3: Total number of possible design variables - analyzed region of aft fuselage stringers, frames and skin panels

feedback loop module and hence are not considered in the next total iteration loop. Indeed, multiple iterations are performed inside this mass penalty estimation approach until no internal loads exceedance appears. In addition, a conservative approach is chosen in which the mass is not reduced at any region of the aft fuselage and the properties of affected elements are scaled up only. The total number of possible design variables based on the analyzed aft fuselage stringers, frames and skin panels is shown in Table 4.3.
The implemented approach for the estimation of mass penalties is purely based on stress thresholds, maximum stress values and internal loads envelopes. In future studies this mass penalty estimation process can be extended. For instance a more sophisticated approach, which uses the global FEM internal loads and the idea of convex hulls to build failure envelopes, is shown in the work [Dha15].

Note again that the main objective of this thesis is the assessment of the impact of an active load alleviation system on the wing and fuselage components. The idea is to make conclusions based on relative values for which the results of a baseline load case set (with de-activated LAF) are compared with the results of the current load case set (with activated LAF). In this sense the effect of cabin pressure and temperature on the fuselage is not taken into account. At this point it is referred to the Master Thesis [Ger14] in which the additional loads due to temperature effects and cabin pressure are analyzed with the support of surrogate models. It is shown that the cabin pressure has a significant effect on the structural reserve factors while the impact of temperature loads is neglectable.
Indeed in future studies the mass penalty estimation approach can be improved for instance by additional static strength or fatigue and damage tolerance criteria.

5 Feedback loop

5.1 Starting approach

The change of stiffness and mass values of the wing structure due to the structural property optimization has to be taken into account in the loads calculation model to start a new loads loop. The values of both quantities need to be extracted from the 3-dimensional FEM model and to be transferred to the 1-dimensional loads stick model. The MDAO process uses two different types of FEM models and each of them is differently discretized based on their purpose of application. In this sense the FEM model for structural optimization is called *model for strength assessment* whereas the FEM model which is used inside the feedback loop is named *model for dynamic analysis*.

A fine discretization is used for the *model for strength assessment* in order to have a correct internal loads distribution as well as an accurate representation for strength assessment. On the other hand a rough but still accurate discretization is used for the *model for dynamic analysis* with the purpose of a correct overall structural stiffness representation which in the end leads to the computation of reliable displacement values. For each component of the aircraft there is one separated *model for strength assessment*, e.g. the wing and fuselage, whereas only one *model for dynamic analysis* for the total aircraft exists. One can conclude that the finer the discretization of the *model for dynamic analysis* the more computationally expensive is the processing of the feedback loop.

The change of stiffness values of the *model for strength assessment* has to be introduced into the *model for dynamic analysis* which is the reference model in the feedback loop process. The transfer of delta stiffness values is based on an *equivalent beam* approach which uses the condensed stiffness matrix of the wing FEM model in order to calculate *equivalent beam* stiffness values. Here *equivalent beam* properties are computed for the wing of the FEM *model for strength assessment* and applied to a chain of beam elements which are modelled along the elastic axis of the wing FEM *model for dynamic analysis*. Thus using the *model for dynamic analysis* as a reference one and adding delta stiffness values at the same time the change of stiffness is considered in the loads calculation model at the end of the feedback loop. The detailed process steps are explained in the coming sections.

In addition, the Grid Point Weight Generator (GPWG) of MSC Patran® is used for the update of the mass values. Here again the delta values are retrieved from the *model for strength assessment* and transferred to the *model for dynamic analysis*.

Both the *equivalent beam* approach and the update of the mass values are explained in more detail in the next sections.

5.2 Main process steps of the feedback loop

In this section the main process steps of the feedback loop which is a main pillar of the total MDAO process are described. Here in particular the locations of both stiffness and mass matrix update processes inside the feedback loop are highlighted. The general program frame to come from the *model for dynamic analysis* to the loads calculation model is provided by Airbus. This Airbus program package is called MDAACE and is here adapted for the objectives of this thesis. Accordingly most of the data that is needed inside the feedback loop is already available with initial settings. This data contains for instance the stiffness and lumped mass distribution, the aircraft geometry definition, the aerodynamic panel model and database for correction, the spline matrix for structural and aerodynamic model connection, the dedicated loads monitoring and cross sections with an appropriate transformation matrix, and a simulation model for the EFCS. In this sense the standard steps like static condensation or modal analysis are described very shortly whereas the own developments regarding stiffness and mass update are explained in more detail.

The main steps of the feedback loop are visualized in the two flow charts, which are shown in Figure B.1 and B.2, and highlight only those steps and information that are mainly affected by the wing structural property optimization. The feedback loop starts with the static condensation, following the approach of Guyan (cf. [Guy65]), of both the wing *model for strength assessment* before and after the structural property optimization. The output of the condensation step is the stiffness matrix $\mathbf{K}_{g1,wing,SMi}$ for each of the wing FEM models . The index g1 indicates that the FEM model is condensed to the grid points of the primary structure. In a next step the change of vertical, lateral and rotational wing stiffness is calculated based on the *equivalent beam* approach. Here the stiffness matrices, predefined grid point coordinates for the chain of beam elements and material properties are mandatory. Normally the latter one is retrieved from the FEM *model for strength assessment* and thus already comprised inside the stiffness matrix. Alternatively one can define different material properties explicitly. The change of wing stiffness is passed to the full FEM aircraft *model for dynamic analysis* and thus the change of wing stiffness is built into the condensed stiffness matrix $\mathbf{K}_{g1,total,DM}$. The update of the lumped masses along the wing due to the structural property optimization is considered in the *model for dynamic analysis* as well. Together with the condensed stiffness matrix $\mathbf{K}_{g1,total,DM}$ the mass data is used for a free-free modal analysis of the aircraft (cf. [Res06], [WC15], [VCO17]).

The modal analysis is performed with the MSC Nastran® solution sequence and provides the modal matrix Φ_i as well as the mass matrix $\mathbf{M}_{g1,i}$ based on the concentrated lumped masses for each mass condition. The modal matrix for each mass condition contains the different mode shapes, which are affected by the structural property optimization, for their corresponding eigenvalues. For the manoeuvre calculation the mass and stiffness matrices are used to compute the SMT values in the trim condition for certain flight points and load factors. If continuous turbulence cases are computed the results of 1g trim manoeuvres together with stiffness and mass information are retrieved first. Based on these pieces of information and applying the modal matrices the gust increments are computed and added to the 1g manoeuvre results in order to come to the total SMT values.

5.3 Update of stiffness values - equivalent beam approach

As mentioned earlier the delta stiffness values resulting from the optimization of the *model for strength assessment* are transferred to the *model for dynamic analysis* by using the *equivalent beam* approach. The main idea of this approach is to approximate the elastic behavior of a beam like structure, e.g. the aircraft wing, through a chain of beam elements. The modeling approach is demonstrated in Figure 5.1.

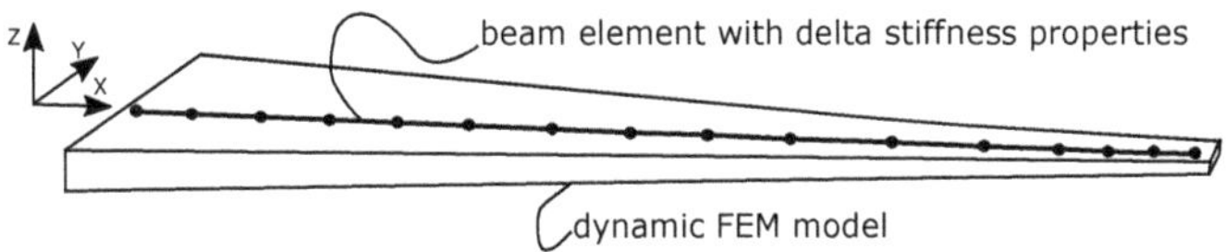

Figure 5.1: Schematic figure of equivalent beam approach

In the paper [WC10] a successful application of an *equivalent beam* model in the frame of fluid structure interaction is presented. In this sense the authors of [WC10] apply the resulting loads after the CFD calculation to the beam elements instead of using a 3-dimensional FEM model. The main argument of the authors for the use of an *equivalent beam* model for the aeroelastic part is the computationally inexpensive analysis plus the time that is saved for the preparation of a 3-dimensional FEM model. In the work [WC10] "the behavior of the structure is predicted by Euler-Bernoulli beam theory, which assumes that the cross-sections remain flat and perpendicular to the beam neutral line. This linear structure model assumes small displacements and rigid cross-sections". The authors assume the Euler-Bernoulli beam theory to be accurate enough to model the flexible wing deformation during the iterative CFD and aeroelastic analysis until an equilibrium condition is achieved. Further in the journal paper [ESA09] and the thesis [Cir11] different methodologies to derive *equivalent beam* properties from a 3-dimensional FEM model are shown.
In case both the *model for strength assessment* and the *model for dynamic analysis* are identically discretized the transfer of delta stiffness values can be achieved by using the condensed stiffness matrix directly. Major advantage of this idea is the higher accuracy due to the couple terms in the off-diagonal entries, which are indeed neglected when using *equivalent beam* properties. However due to the different discretization of both FEM models a chain of beam elements must be used in order to transfer the delta stiffness values to the *model for dynamic analysis*. The *equivalent beam* approach that is used in this thesis is an Airbus internal development. Hence the main assumptions and steps are described only. The first major step is the static condensation of the *model for strength assessment* according to Guyan (c.f. [Guy65]) by using the MSC Nastran® solver. The following calculations and matrix manipulations are all based on this condensed stiffness matrix. Note that the inner and outer wing box are divided into a number of smaller boxes by ribs. Inside these smaller wing boxes RBE3 elements are used, e.g. to apply forces or concentrated mass elements, and whose dependent nodes are connected to the surrounding wing covers. In this sense the locations of the condensation points are chosen in accordance to the coordinates of the RBE3 master nodes. The chain of beam elements is connected to these RBE3 master nodes via RBE2 connections. For each of the n con-

densation grid points, that are placed along the wing elastic axis, the total 6 degrees of freedom are taken into account which yields a symmetric condensed stiffness matrix containing $n\times6$ rows and columns.
The Guyan reduction yields the condensed stiffness matrix considering the condensation grid points that are placed along the elastic axis from the root to the tip of the wing. In the following steps matrix manipulations are applied in order to extract the *equivalent beam* properties from this condensed stiffness matrix. The starting point for these matrix manipulations is the FEM equation for linear static analysis which is shown in equation 5.1. The equation implies that the product of condensed stiffness matrix $\mathbf{K}_{condensed}$ and related displacement vector $\mathbf{u}$ equals the vector of applied forces and moments $\mathbf{f}$.

$$\mathbf{K}_{condensed}\,\mathbf{u} = \mathbf{f} \tag{5.1}$$

In a first step the degrees of freedom at the wing root are fixed, comparable to a clamped cantilever beam, which results in a further reduction by 6 degrees of freedom of the already condensed stiffness matrix. In order to calculate the displacements the condensed and reduced stiffness matrix $\mathbf{K}_{condensed,reduced}$ is inverted (cf. equation 5.2)

$$\mathbf{u} = \mathbf{K}^{-1}_{condensed,reduced}\,\mathbf{f} \tag{5.2}$$

The next step is similar to applying 6 unit forces and moments to the tip of a fixed cantilevered beam successively in order to calculate its displacement values. This step results in the extraction of the last 6 columns from the condensed and reduced stiffness matrix. These columns contain the resulting translations and rotations for each grid point due to the 6 unit loads. When using unit forces and moments in the vector of applied loads $\mathbf{f}_{unit}$ iteratively the extracted columns of the inverted stiffness matrix yield the flexibility matrix $\mathbf{F}_{bb}$ (cf. equation 5.3).

$$\mathbf{u}_{unit} = \mathbf{F}_{bb}\,\mathbf{f}_{unit} \tag{5.3}$$

In order to proceed with the *equivalent beam* approach a chain of beam elements is assumed for which the properties are evaluated iteratively. Therefore starting from the tip the first clamped beam element consists of grid point n and $n-1$. Similar to the approach that is shown in [ESA09] the remaining grid points towards the wing root are also supposed to be clamped. If one refers to the reduced 6×6 stiffness matrix $\mathbf{K}^e_{22}$ of a fixed-free finite beam element (cf. [PHEL09]) as shown in equation 5.4, then its inverse is assumed to be equal to the 6×6 sub-matrix as received for each grid point after the extraction of the displacement vectors from the flexibility matrix $\mathbf{F}_{bb}$.

$$\mathbf{K}^e = \begin{bmatrix} \mathbf{K}^e_{11} & \mathbf{K}^e_{12} \\ \mathbf{K}^e_{21} & \mathbf{K}^e_{22} \end{bmatrix} \tag{5.4}$$

In order to be consistent with the property definition of the global FEM model the proposed *equivalent beam* approach applies an Euler-Bernoulli beam, too. Accordingly the inverse stiffness matrix $(\mathbf{K}_{22}^e)^{-1}$ is shown in equation 5.5. In order to derive the inverse stiffness matrix $(\mathbf{K}_{22}^e)^{-1}$ of each beam element inside the chain it is necessary to proceed as shown in equation 5.6 which is also proposed in a similar way in [Cir11]. In order to be more generic one can label grid $n-1$ as point 1 and grid n as point 2. Then the flexibility matrix values of point 1 are subtracted from those of point 2 because the degrees of freedom at point 1 are supposed to be fixed. Different to the approach that is shown in [Cir11] the rigid body transformation matrix $\mathbf{G}$ (cf. [M+81], p. 102) is applied to evaluate the flexibility matrix of point 1 with respect to point 2 before the subtraction. Note that in this context small displacement values are assumed due to the unit loads. The transformation matrix is also used to evaluate the unit loads from the tip to point 2. Equation 5.6 is applied to all the single beam elements inside the chain.

$$(\mathbf{K}_{22}^e)^{-1} = \begin{bmatrix} \frac{L}{EA} & 0 & 0 & 0 & 0 & 0 \\ 0 & \frac{L^3}{3EI_z} & 0 & 0 & 0 & \frac{L^2}{2EI_z} \\ 0 & 0 & \frac{L^3}{3EI_y} & 0 & -\frac{L^2}{2EI_y} & 0 \\ 0 & 0 & 0 & \frac{L}{GJ} & 0 & 0 \\ 0 & 0 & -\frac{L^2}{2EI_y} & 0 & \frac{L}{EI_y} & 0 \\ 0 & \frac{L^2}{2EI_z} & 0 & 0 & 0 & \frac{L}{EI_z} \end{bmatrix} \tag{5.5}$$

The rigid body transformation matrix $\mathbf{G}$ uses the distance from one point to another whereas an additional transformation from the basic to the local coordinate system is not taken into account.

$$(\mathbf{K}_{22}^e)^{-1} = (\mathbf{u}_2 - \mathbf{G}_{12}\,\mathbf{u}_1)\,(\mathbf{G}_{T2}\,\mathbf{f})^{-1} \tag{5.6}$$

The rigid body transformation matrix which is used to transform the desired quantities from point 1 to 2 is used as described in [M+81] and is shown in the following equation 5.7.

$$\mathbf{G}_{12} = \begin{bmatrix} 1 & 0 & 0 & 0 & (x_{3,2}-x_{3,1}) & -(x_{2,2}-x_{2,1}) \\ 0 & 1 & 0 & -(x_{3,2}-x_{3,1}) & 0 & (x_{1,2}-x_{1,1}) \\ 0 & 0 & 1 & (x_{2,2}-x_{2,1}) & -(x_{1,2}-x_{1,1}) & 0 \\ 0 & 0 & 0 & 1 & 0 & 0 \\ 0 & 0 & 0 & 0 & 1 & 0 \\ 0 & 0 & 0 & 0 & 0 & 1 \end{bmatrix} \tag{5.7}$$

Once the inverse stiffness matrix of the beam element is determined the *equivalent beam* properties EI_y, EI_z and GJ are calculated by comparing these values to the matrix which is shown in equation 5.5. The procedure starting from the static condensation until to the determination of the property values is performed for both the initial and optimized *model for strength assessment*. In this way the delta property values are calculated and applied to the MSC Nastran® PBEAM cards which are used for the chain of beam elements in the *model for dynamic analysis*. Note that for a reduction of the stiffness negative entries are supported for E and G in the MAT1 card of MSC Nastran® version 2005. If the structural property optimization of the *model for strength assessment* leads to a reduction of the wing stiffness then this effect can be considered by using this feature. In

this way the stiffness that is applied to the chain of beam elements is subtracted from the *model for dynamic analysis*
A validation of the proposed *equivalent beam* approach can be found in the appendix C. In order to validate the *equivalent beam* approach one of the vertical balanced manoeuvre cases, which are also used for the design studies, is applied to the different forms of the wing structure. Here the wing FEM *model for strength assessment*, the condensed form of this FEM according to Guyan and the representation by the chain of beams with its *equivalent beam* properties are used. The displacement values in z-direction are compared to each other for the purpose of validation. Further in order to have a better understanding of the influence of upper and lower wing cover properties on the resulting *equivalent beam* stiffness values a small variation analysis is performed. The results are shown in the appendix D.

5.4 Update of mass values - grid point weight generator

The total mass contributions are considered when working with the *model for dynamic analysis* in the feedback loop. Here in addition to the structural mass of the wing box the contributions of all slats, flaps, ailerons, systems installation and equipment, operators items and unusable fuel are included. All these contributions lead to the OWE of the wing. Note that the value of the unusable fuel mass is assumed to be constant. It is not accessible by the engines due to its location in pockets within the wing box between structural parts and in the pipes. Different total mass conditions are reached by adding the corresponding amount of usable fuel to the OWE. The same idea applies to the fuselage section but here the main focus is on the wing. Hence the update of the wing mass due to the structural optimization is implemented in the feedback loop only.
As described in the PhD thesis [Öst06] the total aircraft mass m_{tot} is the summation of partial masses (cf. equation 5.8). The mass of the primary structure is predicted based on a 3-dimensional geometry definition which is used for the structural sizing and adapted for the FEM model. In this way one receives the mass of the primary structure m_{FEM} which is then used as a baseline. In order to consider additional structural masses which are not taken into account in the FEM model, i.e. rivets or manhole covers, the value of m_{FEM} can be factorized. A second term m_{add} is added respectively for each FE and includes all other mass contributions that are independent from the structural sizing.
In the research [Rie13], in which the mass estimation process is mainly based on the work that is shown in [Öst06], an overview of different approaches and tools is provided. Here statistical methods are mentioned for which the mass is estimated by comparing to already existing aircraft components. Further analytical procedures as well as methods for the preliminary design of the aircraft primary structure are reviewed for mass estimation purposes. Another way to estimate component masses based on the FEM model and additional empirical factors is described in [WSG12]. Furthermore in the conference paper [KLG95] another approach, that is purely based on classical theories instead of FEM models, for the wing primary structure mass estimation is presented.

$$m_{tot} = m_{FEM} + m_{add} \tag{5.8}$$

The developed approach that is presented in this PhD thesis is based on concentrated lumped masses, i.e. OWE with additional fuel, based on the FE discretization of the *model for dynamic analysis.* The condensed 1-dimensional mass model is realized by attaching CONM2 cards to the associated grid points using RBE2 elements. A set of CONM2 cards for different mass cases with their related center of gravity and moments of inertia is provided as an input. The idea is to use the total mass values that are given for each CONM2 card as a starting point and assume an additive behavior as proposed in [Öst06]. In this sense concentrated delta masses are calculated based on the structural optimization of the *model for strength assessment* and applied to the CONM2 cards of the *model for dynamic analysis.*
For this purpose a PCL script is written that launches the GPWG of MSC Patran® in order to calculate concentrated mass values along the wing span for the targeted elements and regions. The basic equations behind the GPWG are provided in the appendix E. As shown in Figure 5.2 the FEM *model for strength assessment* of the wing is split into the same number of slices as concentrated masses are existing for the *model for dynamic analysis.* The segregation lines along the wing are set parallel to the streamlines which means that single elements are split as well and hence their mass contributions have to be distributed to the associated slices. Since the structural optimization is performed for the wing upper and lower covers only the GPWG calculates the mass values for Quad4, Tria3 and Bar elements inside each slice where each value is distributed equally to the element grid points. The summation of the grid point masses inside each slice results in the concentrated mass value.

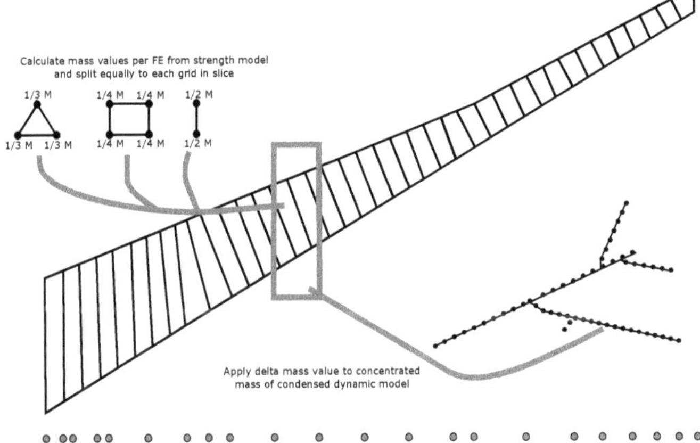

Figure 5.2: Schematic figure of concentrated mass evaluation

The concentrated masses are evaluated for the initial and optimized FEM *model for strength assessment* of the wing and the resulting delta values are added to the total mass value of each CONM2 card (cf. equation 5.9). Since the main objective is to assess sensitivities only delta mass values are considered inside the MDAO process as it is proposed in [WSG12], too.

$$m_{i,tot,new} = m_{i,tot,old} + (m_{i,FEM,2} - m_{i,FEM,1}) \tag{5.9}$$

In addition to the mass values also the main ($I_{m,11}$, $I_{m,22}$, $I_{m,33}$) and secondary ($I_{m,21}$, $I_{m,31}$, $I_{m,32}$) mass moments of inertia of each CONM2 entry are updated. Here the assumption is that the first order radius of gyration r_g , which is a characteristic value for the inertia properties (cf. [Ard06]), stays constant when the change of structural properties is small (cf. equation 5.10).

$$r_{g,new} = r_{g,old} \tag{5.10}$$

The terms for the new and old radius of gyration in equation 5.10 are replaced by their explicit forms which then yields equation 5.11.

$$\sqrt{\frac{I_{m,i,new}}{m_{i,tot,new}}} = \sqrt{\frac{I_{m,i,old}}{m_{i,tot,old}}} \tag{5.11}$$

Based on equation 5.11 the relation for the new inertia values is derived as follows:

$$I_{m,i,new} = I_{m,i,old} \frac{m_{i,tot,new}}{m_{i,tot,old}} \tag{5.12}$$

Another assumption that is made for the update of the concentrated mass values regards the center of gravity (cf. equation 5.13). The structural properties of the inner and outer wing box covers are optimized only. Assuming a constant density value for this optimization region its weight accounts around 30% of the wing OWE. With respect to this scale and performing small optimization steps the center of gravity for each CONM2 entry is assumed to be unchanged.

$$CG_{i,new} \approx CG_{i,old} \tag{5.13}$$

This assumption is also confirmed in the forthcoming chapter 6 in which the CG coordinates $x_{CG,i}$, $y_{CG,i}$ and $z_{CG,i}$ are re-calculated for the updated structural model according to the equations 5.14, 5.15 and 5.16. Here for each slice the sum of the product of grid point mass values and corresponding coordinate is divided by the total mass value.

$$x_{CG,i} = \frac{\sum_{i=1}^{n} m_i\, x_i}{m_{tot,i}} \tag{5.14}$$

$$y_{CG,i} = \frac{\sum_{i=1}^{n} m_i\, y_i}{m_{tot,i}} \tag{5.15}$$

$$z_{CG,i} = \frac{\sum_{i=1}^{n} m_i\, z_i}{m_{tot,i}} \tag{5.16}$$

In this way the mass values are updated and the new entries are considered in the mass matrix for modal analysis and the calculation of gust loads.

5.5 Global convergence criteria

A convergence criterion is needed in order to assess the end of the global MDAO loop before a new loads calculation starts. Note that for this purpose the updated fuselage properties are not considered for convergence considerations because on the one hand only mass penalties are estimated and on the other hand related modifications are not transferred to the forthcoming iteration loop. Therefore the change of mass and stiffness along the wing component is the main criterion. The former is assessed through the concentrated masses and the latter using the *equivalent beam* stiffness from one iteration loop to another.
The boundary value for mass convergence is parametrized inside the tool environment and the process stops when the change of each concentrated mass element is less than or equal to this value. A first appropriate boundary value that is based on multiple performed simulations is given in equation 5.17. Here the structural property optimization of the wing covers converges with a maximum number of 10 iteration loops of the total MDAO process chain when applying this mass convergence value.

$$\varepsilon_m \leq 0.1\,kg \tag{5.17}$$

Using the current tool settings the process ends when the mass convergence criterion is reached. Indeed, the stiffness criterion is also implemented inside the tool environment but however is not actively used in the current version. In order to assess the change of overall wing stiffness more profoundly a modal analysis has to be performed for instance. Thus the impact of stiffness changes on the mode shapes and frequencies can be evaluated and doing so the stiffness criterion can be used as a constraint more effectively. Due to the limited time frame of this thesis the results of the modal analysis, which is part of the feedback loop, are not actively built in as part of a convergence criterion.

5.6 Impact of wing stiffness changes on the jig and flight shape

An important issue that is unfortunately not covered in this thesis is the influence of different wing stiffness distributions on the jig shape and the related flight shape of the wing component. The jig shape is the undeformed wing shape without any inertial or aerodynamic forces applied (cf. [WC15]). On the other hand the flight shape is the so-called "optimal shape" under deformation during 1g cruise flight in a state of aeroelastic equilibrium and for a predefined design point (cf. [WC15], [AMB03], [Hür10]). This design point for an optimal cruise flight consists of total aircraft mass, mach number and altitude. For a known jig shape the flight shape can be found by computing the structural deformation and the resulting aerodynamic loads in an iterative process until convergence to an aeroelastic equilibrium (cf. [AMB03], [Hür10]). In the case of a given flight shape in the conference paper [KM14] a method to come to the corresponding jig shape by applying an inverse design procedure is proposed. The resulting jig shape can then be used for further aerostructural optimization. The structural property optimization that is performed in this thesis has an impact on the wing stiffness and thus the original flight shape is also

affected as stated in [JPM10]. Due to its different bending and torsional behavior the lift distribution along the wing is not optimal anymore in terms of aerodynamic aspects, i.e. drag minimization, as it is the case with the original flight shape. In industrial practice the jig shape is provided and then the structural elements like stiffeners, skin panels, ribs and spars are sized such that the structural deformation under 1g loading matches the flight shape (cf. [MAR01]).
The effect of stiffness changes on the flight shape for a given jig shape in terms of aerodynamic efficiency is not considered in this thesis and hence it is a step forward when this aspect is taken into account in a future study.

6 Design studies

In this chapter different design studies are performed using the total MDAO process that includes the load calculation, LAF parameter variation, structural assessment and the feedback loop. For this purpose the design space, boundary values, initial properties and settings as described in the previous chapters are used. In this sense the vertical balanced manoeuvre and continuous turbulence cases are calculated with a de-activated LAF and then again with an activated LAF. Further the structural properties of the wing covers are modified once per total iteration loop and then send back to the loads calculation model via feedback loop. The MDAO process runs until mass convergence of the wing is reached or otherwise is stopped after a predefined maximum number of 10 total iteration loops. Afterwards the mass penalty of the aft fuselage region is estimated based on both the chosen LAF parameter settings and the converged wing cover properties.
Here the influence of the load alleviation function on the calculated loads is assessed by performing a sensitivity study first. For this purpose flight manoeuvre cases with different load factor values and LAF parameter settings are calculated whereupon the impact of the latter ones on the SMT loads and the normalized lift distribution is evaluated.
As stated before the used wing and fuselage models, which are provided by Airbus, are already statically sized based on a certain set of load cases and failure criteria. Here the applied load cases and failure criteria are a subset of the original ones only. Hence the idea is to make a relative quantification of the mass reductions and mass penalties due to the LAF parameter variations with respect to a generated baseline model. For this kind of baseline model generation the MDAO process is run with a de-activated LAF and the chosen list of load cases until mass convergence of the analyzed wing cover regions. The generated wing baseline model is then the starting point for the assessment of possible mass reductions due to an activated LAF and related mass penalties for the fuselage section.
For the design studies those regions which are affected the most by the changes in loads due to the LAF are selected only. Figure 6.1 shows on a generic long range aircraft FEM model those parts (exploded view) of the upper and lower wing covers that are optimized. The highest changes in the element stresses and forces due to the LAF appear in these regions. Further the landing gear attachments are located close to the wing root and hence these parts of the wing covers are primarily sized by ground loads. The lack of ground loads in this design study can lead to significantly high mass reductions (when optimizing with flight loads only) and in this sense these regions are excluded from the optimization. In addition, the property values of the wing covers that are located close to the wing tip are already small and close to the lower boundary values. Hence these regions are excluded from the optimization, too. Figure 6.2 shows the fuselage analysis region (exploded view) for the purpose of mass penalty estimation on a generic long range aircraft FEM model. The upper and lower parts of the aft fuselage shell are chosen only because the highest changes in the element stress and force values appear in these regions. In addition, on the one hand vertical manoeuvre and continuous turbulence cases are calculated only and on the other hand the additional force that is introduced into the

Wing covers			
Finite element	**Number of elements**	**Design variable**	**Number of design variab**
Quad4	766	t	766
Tria3	16	t	16
Bar	732	$L_1, ..., L_6$	4392
Aft fuselage shell			
Finite element	**Number of elements**	**Design variable**	**Number of design variab**
Quad4	1390	t	1390
Tria3	74	t	74
Bar	1423	L_5	1423
Rod	1084	A	1084

Table 6.1: Number of design variables - wing covers and aft fuselage shell

fuselage as a consequence of the LAF parameter variation and the resulting pitch moment compensation is also vertical. In this sense the resulting number of design variables is listed in Table 6.1. Here the number of design variables for each type of finite element is listed for the chosen analysis region and is the product of the number of elements and the number of design variables of each finite element.

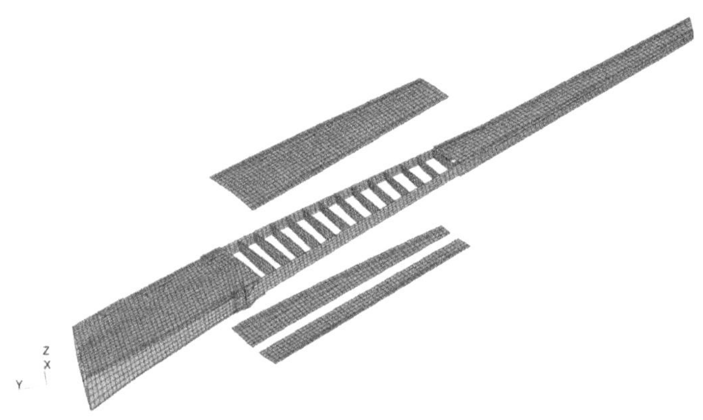

Figure 6.1: Wing optimization region - shown on a generic long range aircraft FEM model

The design studies are performed in 2 steps. In order to derive some key conclusions in less computation time 16 load cases are applied in the first step only. These 16 load cases build a representative subset of the total 536 load cases (cf. chapter 3.1) which are used here. The conclusions regard in particular the convergence of the structural property optimization algorithm and the impact of the LAF on the structural components. In the second step the total 536 load cases are applied in order to confirm the derived conclusions. In addition, the convergence of the FSD optimization algorithm is shown by using two different sets of initial values for the wing cover skin and stringer properties. The first set of initial property values is retrieved from the wing FEM model as it is provided by Airbus. The second set of initial property values is given by scaling the skin and stringer property values to predefined minimum values. The set of initial property values for the aft fuselage shell is retrieved from the FEM model as it is provided by Airbus.

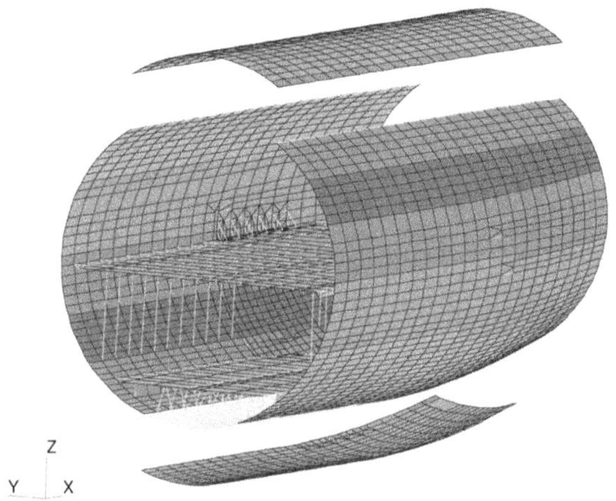

Figure 6.2: Fuselage analysis region - shown on a generic long range aircraft FEM model

In the forthcoming sections it is shown that the implemented FSD optimization algorithm is partially validated. Note that a complete validation of the implemented method is not trivial due to the lack of the complete set of sizing load cases and failure criteria.

6.1 Influence of the load alleviation function on the calculated flight loads

One major advantage of the developed MDAO process is its capability to quantify and plot parameters from different disciplines like the flight loads calculation or the structural sizing. In this sense before the impact of the load alleviation function on the wing and fuselage components is shown here the impact on the calculated flight loads is demonstrated first. For this purpose a sensitivity study is performed in which vertical flight manoeuvre cases are calculated for different load factor values and with changing LAF parameter settings. The impact on the results is assessed based on SMT values and the normalized lift distribution.
In this sense Figure 6.3 shows the effect of different LAF parameter settings on the normalized lift distribution of a vertical flight manoeuvre ($n_z = 2.5g$) along the normalized wing semispan η. Here the lift values are normalized to the maximum value of the lift distribution that results when the LAF is inactive. It can be seen that the center of lift is shifted inboard (shown in the figure as ΔCoL) when the inner and outer ailerons are

deflected upwards by the value $\delta_{in,out}$ = -15°. The shift of the center of lift inboard is equivalent to a reduction of the wing root bending moment since the lift at the outer wing region is destroyed due to the aileron upwards deflection. Note that although the shape of the lift curve changes still its integral stays the same. In addition, it is shown that the outer aileron has more impact on the lift curve than the inner aileron. This is shown by deflecting the outer aileron upwards by the value δ_{out} = -15° and keeping the inner aileron undeflected which then yields a lift curve that nearly matches the one that is generated by deflecting both ailerons.

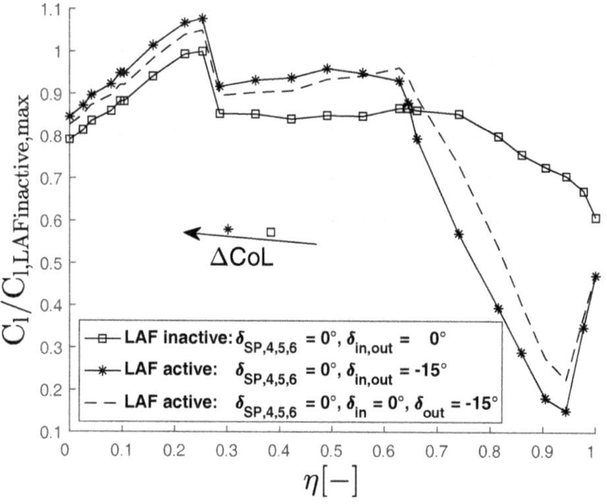

Figure 6.3: Effect of inner and outer aileron upwards deflection on the normalized lift distribution of a vertical flight manoeuvre: $n_z = 2.5g$

Indeed, a load alleviation is connected with an additional pitching moment originating by aileron and spoiler deflections. Figure 6.4 shows the effect of different LAF parameter settings on the normalized pitching moment coefficient C_m of a vertical flight manoeuvre ($n_z = 2.5g$) along the normalized wing semispan η. Especially in those regions where the ailerons are installed it is shown that upwards deflections of the ailerons lead to an increase of the pitching moment coefficient. As shown in equation 6.1 the pitching moment coefficient is defined as the ratio of the pitching moment M and the product of the dynamic pressure q, the wing area S_w and the chord length c (cf. [WC15]).

$$C_m = \frac{M}{q\, S_w\, c} \tag{6.1}$$

Here the dynamic pressure is calculated using the air density and the air speed and is shown in equation 6.2:

$$q = \frac{1}{2}\rho V^2 \tag{6.2}$$

Indeed, the pitching moment is affected by the upwards and downwards deflections of the wing control surfaces and as stated in [WC15] "there is also the potentially disastrous phenomenon of divergence to consider, where the wing twist can increase without limit when the aerodynamic pitching moment on the wing due to twist exceeds the structural restoring moment". Note that the change of the lift distribution or the pitching moment coefficient can be assessed as shown but aeroelastic phenomenons like divergence are not investigated here.

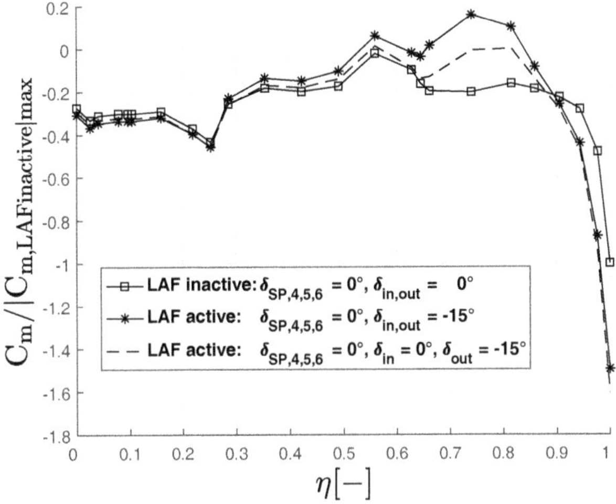

Figure 6.4: Effect of inner and outer aileron upwards deflection on the normalized pitching moment coefficient of a vertical flight manoeuvre: $n_z = 2.5g$

Figure 6.5 shows the percentage change of the center of lift ΔCoL dependent on the inner and outer aileron deflection angles for vertical manoeuvres with different load factors. Here each manoeuvre (2.5g, 1g and -1g) is re-calculated with different values for the inner and outer aileron deflection angles whereupon in each case the change of the center of lift is evaluated. The value ΔCoL can be used as a supporting parameter for better understanding in future studies when aerodynamic aspects are investigated in more detail. For instance in the case of a vertical flight manoeuvre with positive load factor and a backward swept wing the center of lift is shifted inboard and backward due to the load alleviation function. In this sense in Figure 6.5 the value of ΔCoL shows the shift inboard along the wing y-axis (root to tip) which affects the vertical bending moment M_x the most. It can be seen that for the 2.5g manoeuvre even when applying both ailerons the shift of the center of lift converges to a value of about 16% for aileron deflection angles which are lower than -10°. With respect to handling quality targets (cf. chapter 3.3.2) this argument can be used before considering higher or lower aileron deflection angles. Using the same aileron deflection angles for the 1g case the center of lift is shifted more inboard (delta value of around 39%) than it is done for the 2.5g case. Also there is no convergence behavior for the ΔCoL value as it appears for the high load factor manoeuvre. Note that the lift for the steady flight case is much lower than for the pull-up manoeuvre and thus its wing deflection and bending moment are smaller, too. In the case of an activated LAF for negative load factor manoeuvres the ailerons are deflected downwards. As can be seen in Figure 6.5 the behavior of the -1g case is similar to the one of the steady flight manoeuvre. One interesting point for this push-down manoeuvre example is the change from negative to positive lift at the wing tip region starting from 9° downwards deflection.

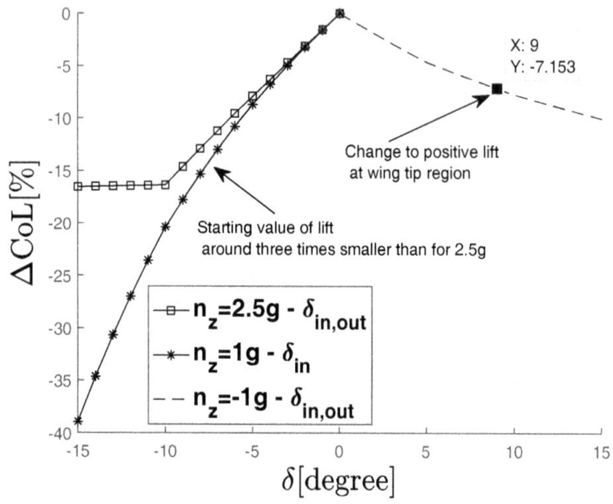

Figure 6.5: Percentage change of the center of lift dependent on the inner/outer aileron deflection angles for vertical manoeuvres with different load factors

A similar behavior can be seen for the related reduction of the wing root bending moment $M_{x,root}$ and torque $M_{y,root}$ in terms of SMT values as it is shown in Figure 6.6 and Figure 6.7. Here again the percentage change of the wing root bending moment and torque is shown dependent on the inner and outer aileron deflection angles for vertical manoeuvres with different load factors. Qualitatively the changes of the wing root bending moment and torque are similar to each other. Only in the case of the -1g manoeuvre the percentage reduction of the torque at the wing root $M_{y,root}$ is more than twice higher compared to the other cases.

The effect of the inner and outer aileron deflections on the normalized lift distribution of a steady flight case and a push-down manoeuvre is shown in the appendix in Figure F.1 and Figure F.2. Similar conclusions can be drawn from these two figures as it is done with Figure 6.3 that shows the lift distribution of the high load factor ($n_z = 2.5g$) pull-up manoeuvre.

The alteration of the lift distribution which leads to a reduction of both the wing root bending moment and torque demonstrates the benefit of the load alleviation function on the calculated flight loads. Indeed, when the analysis scope increases for instance to disciplines like handling qualities or the structural sizing of affected components (wing and fuselage) then appropriate limitations for the control surface deflection angles need to be assessed in order to keep possible disadvantages on quantities of the considered disciplines limited.

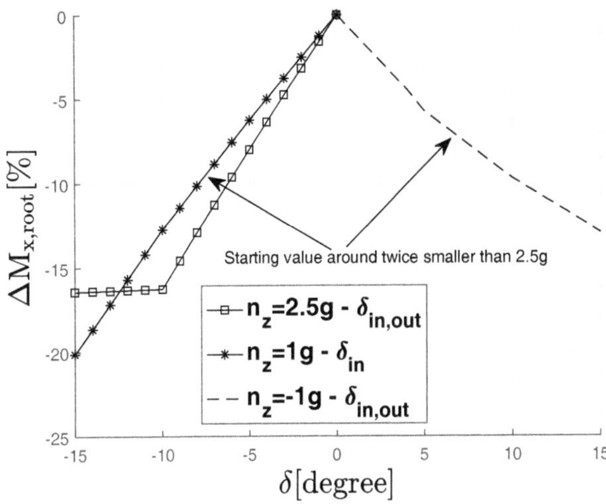

Figure 6.6: Percentage change of the wing root bending moment $M_{x,root}$ dependent on the inner/outer aileron deflection angles for vertical manoeuvres with different load factors

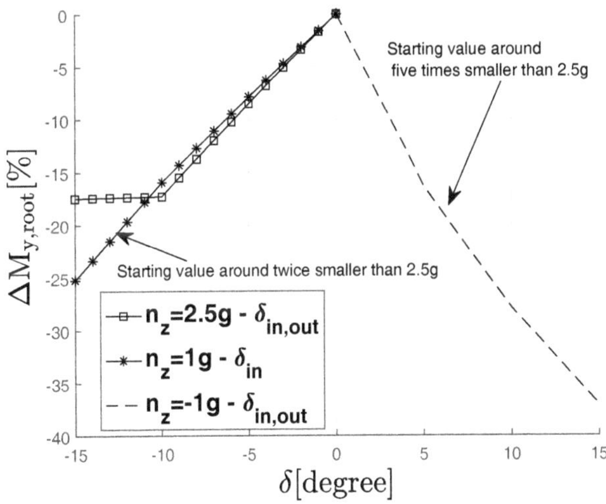

Figure 6.7: Percentage change of the wing root torque $M_{y,root}$ dependent on the inner/outer aileron deflection angles for vertical manoeuvres with different load factors

6.2 First design study with reduced load case set

In this section design studies are performed with a reduced load case set which consists of 16 load cases that build a representative subset of the total 536 load cases (cf. chapter 3.1). The idea is to draw some key conclusions regarding the convergence of the structural property optimization algorithm and the impact of the LAF on the wing and fuselage components in less computation time. The main characteristics of these 16 load cases are as follows:

- 12 vertical flight manoeuvres with positive load factors only (2g, 2.5g)
- 4 continuous turbulence cases which are calculated with positive gust increments only
- Load cases with extended and retracted speedbrakes
- Load cases with zero and maximum thrust
- One single mass case only: MZFW + aft CG
- Load cases calculated with cruise and dive speed
- Load cases calculated with 3 different altitude values

The wing upper cover is predominantly sized in compression and the lower cover is sized in tension. Hence the idea is to consider positive load factor values for this reduced load case set only because the resulting wing upwards bending generates this kind of distribution of tension/compression loads among the wing covers. The mass distribution of the single mass case that is used for the analysis yields an aft CG. Especially for this kind of mass distribution the force increase at the aft fuselage section due to the pitch compensation is high.

In this sense the wing structural baseline model is generated first in order to show the convergence behavior of the structural property optimization algorithms. Here both the implemented stress ratio (cf. chapter 4.4.1) and reserve factor (cf. chapter 4.4.2) based FSD algorithms are applied to the same wing FEM model separately. For this purpose two different sets of initial skin and stringer properties are used. The first set of initial property values is retrieved from the wing FEM model as it is provided by Airbus. The second set of initial property values is given by scaling the skin and stringer property values to predefined minimum values. The set of initial property values for the aft fuselage shell is retrieved from the FEM model as it is provided by Airbus. In the second step the converged wing baseline model is used as a starting point in order to assess the influence of the LAF on both the wing and fuselage component.

Figure 6.8 shows the convergence behavior of the normalized wing cover mass $m_{cov,wing}$ for the two different initial property sets and FSD algorithms. Each data point represents the mass value after one iteration step of the structural property optimization. Here the value of $m_{cov,wing}$ considers the mass of the optimization region that is shown in Figure 6.1. The values are normalized to the maximum mass value $m_{cov,wing,max}$ which is the mass of the optimization region in its initial state. For the baseline generation the initial state is the one of the wing FEM model as it is provided by Airbus. The four simulation runs confirm the convergence behavior of both implemented FSD algorithms. Although different initial property sets are used as a starting point the mass values converge for each of the implemented FSD algorithms. The two implemented FSD algorithms are based on different failure criteria and hence the structural properties converge to two different sets of values accordingly. Using the stress ratio based FSD optimization algorithm the average mass reduction is 8.56%. The final mass values of both simulations that use the stress ratio based FSD algorithm (each simulation with a different set of initial properties) differ 0.85% from this mean value. On the other hand the RF based FSD optimization algorithm yields an average mass reduction of 13.86% in which the final mass values of both simulations differ 0.6% from this mean value.

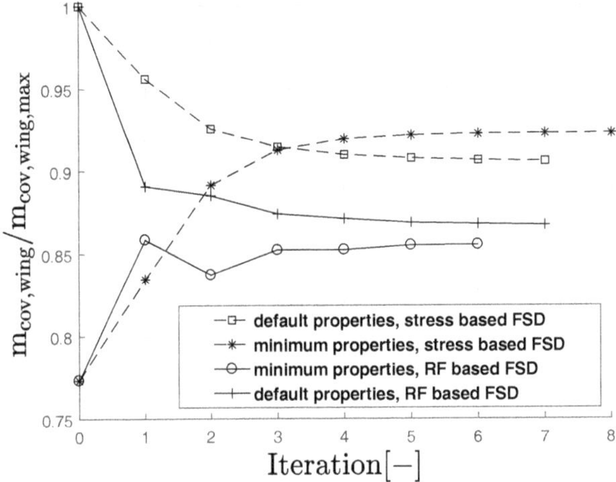

Figure 6.8: Wing structural baseline generation with 16 load cases - convergence behavior of normalized wing cover mass $m_{cov,wing}$ with both two different initial property sets and FSD algorithms

The structural mass of the wing baseline model which is generated by applying the stress ratio based FSD algorithm is significantly higher than the mass of the baseline model that is generated by applying the RF value based FSD algorithm. Indeed, the more sophisticated failure criteria are used the more proper the structure is sized which then leads to a more accurate assessment of the structural mass. Further the distribution of tension/compression loads among the wing covers and the fact that different failure criteria need to be considered accordingly is better realized with the RF based FSD algorithm. The stress ratio based FSD algorithm is based on a single failure criterion only and hence the structural properties of the upper and lower wing covers are modified accordingly.

One drawback when applying the stress ratio based FSD algorithm is the separate modification of the skin and stringer properties. The algorithm changes the properties of each element based on its stress value without taking into account neighboring elements. On the other hand the structural sizing software calculates the RF values for each stiffened panel. Hence the RF value based FSD algorithm modifies the skin and stringer properties according to this stiffened panel assumption. **Based on these aspects the RF value based FSD algorithm is used for the following design studies only.**

This first design study with a reduced load case set shows that with each iteration loop the structural properties are successively modified such that the minimum RF value converges to a value of $RF_{min} = 1$. In addition, it is shown that the upper cover is dominantly sized by the flexural wrinkling failure criterion and the lower cover by the net tension criterion. In this study the sizing load case is a vertical continuous turbulence case with extended speedbrakes.

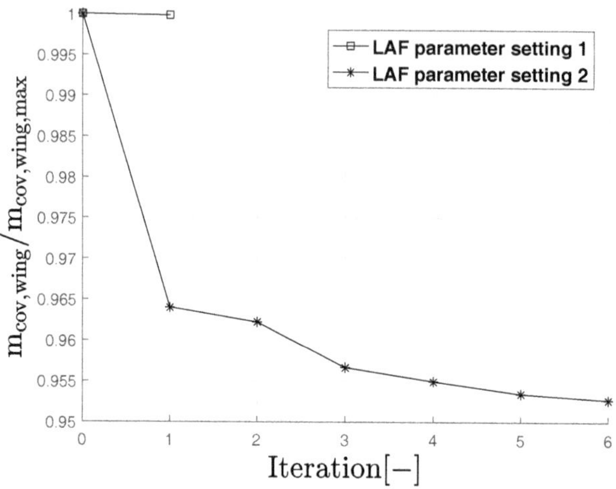

Figure 6.9: Wing structural property optimization with 16 load cases - reduction of normalized wing cover mass $m_{cov,wing}$ due to the LAF shown for two different LAF parameter settings

The next step is to use the converged wing baseline model that is generated with the RF based FSD algorithm as a starting point in order to assess the influence of the LAF on both the wing and fuselage component. For this first design study two different LAF parameter settings are used in which the control surface deflection angles are chosen as described in subsection 3.3.2 (cf. Figure 3.9 and Table 3.2). The control surface deflection angles are the same for both LAF parameter settings but the trigger thresholds for the activation of the LAF are chosen differently. In Figure 6.9 the reduction of the normalized wing cover mass $m_{cov,wing}$ due to the LAF is shown for the two different LAF parameter settings. The value of the wing cover mass $m_{cov,wing}$ refers again to the mass of the optimization region but in this case the values are normalized to the mass of the generated wing baseline model. One can see that for the first LAF parameter setting the achieved mass reduction is neglectable. The trigger thresholds of the first LAF parameter setting are defined as described in subsection 3.3.2 and include threshold values for the vertical load factor n_z and the calibrated airspeed V_{CAS}. As mentioned before the mass trigger is not investigated in this thesis due to the limited time frame. Unfortunately for this first design study with 16 load cases the first LAF parameter setting with all its triggers leads to a new ranking of the sizing load cases. Hence the reduction of the wing structural mass is neglectable because the sizing is performed with those load cases that are not affected by the LAF. Therefore the trigger that is based on the vertical load factor and the one that is based on the calibrated airspeed are both de-activated for the second LAF parameter setting in order to achieve a significant mass reduction of the wing due to the LAF and to perform a first assessment. Using this second LAF parameter setting the wing cover structural mass is reduced by 4.53%. In addition, Figure 6.10 shows the reduction of the normalized wing component mass values due to the LAF for the two different LAF parameter settings. Each component mass value is normalized to the mass of the generated wing baseline model. In this way the contribution of each component to the normalized wing cover mass is shown. In accordance with the change of the wing cover mass value here again the achieved mass reduction of the components for the first LAF parameter setting is neglectable.

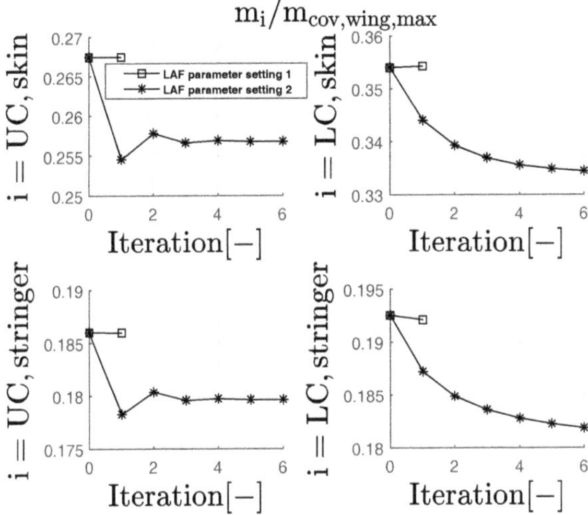

Figure 6.10: Wing structural property optimization with 16 load cases - reduction of normalized wing component mass values due to the LAF shown for two different LAF parameter settings

The lower cover skin panels provide with 35.4% the highest contribution to the mass of the complete wing optimization region. Using the second LAF parameter setting the mass of the lower cover skin panels is reduced by 2%. It follows a mass reduction of 0.8% for the upper cover skin panels, 1% for the lower cover stringers and 0.6% for the upper cover stringers. It is shown that the load alleviation function has more impact on the lower cover components because the mass reduction of the lower cover components is nearly twice the mass reduction of the upper cover components. Note again that here positive load factor values are used only and in addition the upper cover stiffened panels are dominantly sized by the flexural wrinkling criterion and the lower cover ones by the net tension criterion.

The fuselage mass penalty estimation is performed with the second LAF parameter setting only. The significant mass reduction of the wing covers that is achieved when applying the second LAF parameter setting is a proper starting point for the mass penalty estimation of the aft fuselage section. As described in section 4.5 the mass penalty estimation approach is based on internal loads exceedance values for the skin panels, stringers and frames. Here the reduced load case set with 16 load cases is applied to the fuselage global FEM model in order to perform a first assessment. This first design study shows that some mandatory adaptations have to be made to the implemented mass penalty estimation approach. In this sense Figure 6.11 shows the normalized stress values of each Quad4 element of the analyzed aft fuselage region for the 16 load cases and with de-activated LAF. Each stress value is normalized to the allowable stress value σ_{allow} which is set as the yield strength value $R_{p0.2}$ (cf. Table A.1). One can see (plot on the top left) that the normalized von Mises stress $\sigma_{vonMises}$ with values below 0.5 does not get close to the allowable stress value. Note that in the implemented approach the yield strength value is set as a threshold that needs to be passed so that the algorithm applies the internal loads exceedance calculation to the dedicated finite elements. The aft fuselage upper and lower shells are dominantly sized by ground loads which are not considered here. Hence the first mandatory adaptation to the implemented mass penalty estimation approach is the de-activation of the stress threshold which uses the yield strength value. In this way a relative assessment of possible mass penalties due to an activated LAF can be performed although the previously required stress level is not reached.

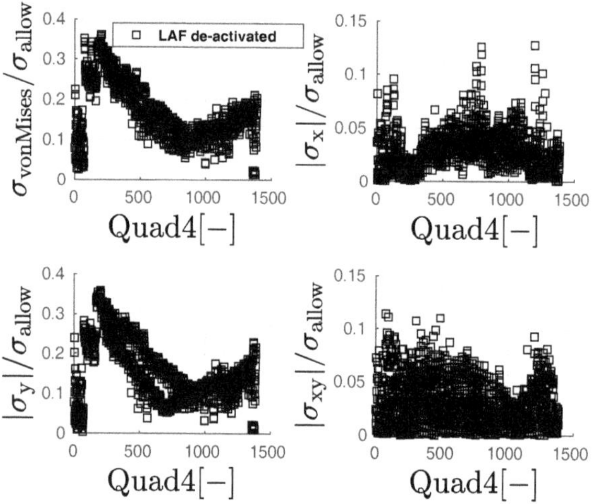

Figure 6.11: Fuselage mass penalty estimation with 16 load cases - normalized stress values of each Quad4 element of the analyzed region shown for load cases with de-activated LAF

In addition Figure 6.11 shows that the axial stress σ_y is more dominant than the circumferential stress σ_x or the shear stress σ_{xy}. Indeed, the dominance of the axial stress values for the analyzed aft fuselage upper and lower shells is related to the applied vertical load cases and the generated tension/compression loads. This dominance of just one internal loads quantity over the other two quantities is a disadvantage when working with 2-dimensional convex hulls because the convex hull that is built appears to be narrow and stretched. In this way the mass penalty estimation approach for the skin panels can lead to misinterpretations of the stress changes at the aft fuselage section which occur due to the elevator demand which is a reaction to the activated LAF. Hence two more adaptations are made to the implemented mass penalty estimation approach. First an additional check for stress increase is implemented. In this way the 2-dimensional convex hull is built for those skin panels (Quad4 elements) only whose resulting von Mises stress is increased by an activated LAF. Those skin panels whose stress values are reduced by an activated LAF are excluded from the mass penalty estimation. Figure 6.12 shows the normalized delta von Mises stress value versus the calculated internal loads exceedance value for each skin panel (one skin panel is modeled with one Quad4 element) of the analyzed aft fuselage region. The delta stress value for each skin panel is calculated with the resulting von Mises stress when applying the reduced load case set with de-activated LAF and then again with activated LAF. Each data point shows the result for one Quad4 element and is normalized to the maximum absolute delta stress value.

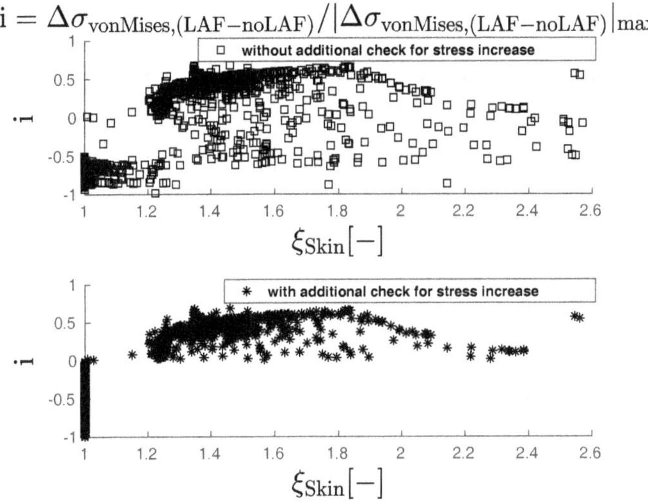

Figure 6.12: Fuselage mass penalty estimation with 16 load cases - normalized delta von Mises stress value $\Delta\sigma_{vonMises,(LAF-noLAF)}$ versus internal loads exceedance value ξ_{Skin} for analyzed aft fuselage skin elements

The plot at the top of Figure 6.12 shows the relation between the change of the von Mises stress and the calculated internal loads exceedance value without the additional check for stress increase. One can see that even for decreasing stress values significant internal loads exceedance values based on the 2-dimensional convex hull are calculated. An increase of the structural properties for elements whose von Mises stress is decreased is avoided by the implemented check for stress increase. The plot at the bottom of Figure 6.12 shows the impact of this additional check for stress increase. Further on the bottom plot it can be seen that even for small stress increases high internal loads exceedance values are calculated which is another drawback when applying the 2-dimensional convex hull. At this stage the algorithm based on the convex hull is used although it gives inconsistent results. Finally Figure 6.13 shows the changes of the normalized component mass values of the analyzed aft fuselage region for the second LAF parameter setting. Here each component mass value is normalized to the initial total mass value of the analyzed aft fuselage region. One can see that even after 10 iterations the mass value does not converge for any of the components. Here again especially the mass increase of the upper and lower skin panels which is calculated based on the 2-dimensional convex hull yields the highest portion of the total mass increase. Hence the remaining adaptation to the implemented mass penalty estimation approach is the limitation to just one iteration loop for the forthcoming design studies.

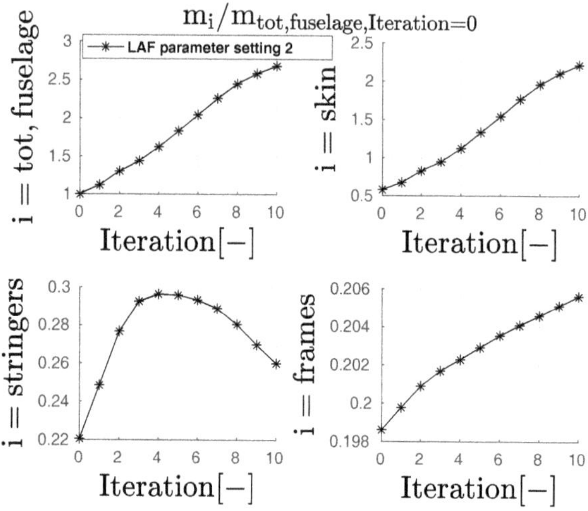

Figure 6.13: Fuselage mass penalty estimation with 16 load cases - changes of normalized component mass values of the analyzed aft fuselage region shown for LAF parameter setting 2

The conclusions that are drawn from this first design study are summarized as follows:

- Wing structural property optimization:
  - ◇ The stress value based FSD algorithm is not used for the forthcoming design studies but the RF value based FSD algorithm is applied only
- Aft fuselage mass penalty estimation:
  - ◇ The stress threshold that is based on the yield strength value is de-activated because the reached stress level is far below this allowable stress value
  - ◇ An additional check for stress increase is implemented which triggers the 2-dimensional convex hull for the mass penalty estimation of the skin panels when the von Mises stress is increased only
  - ◇ Only one iteration loop of the mass penalty estimation is performed because at this stage the implemented algorithm does not lead to a mass convergence

6.3 Design study with complete load case set

In this section the complete load case set with its 536 load cases is applied to the wing and fuselage components. Here the adaptations to the implemented algorithms as described in the previous section are taken into account. In the same way as in the previous section a structural wing baseline model is generated first. This baseline model is then used as a starting point for further assessment of the impact of the LAF on both the wing and aft fuselage component.

6.3.1 Structural baseline generation

Figure 6.14 shows the evolution of the normalized mass of the wing cover optimization region for 6 complete iteration loops of the MDAO process. Each iteration step shows the mass values after the RF value based FSD algorithm is applied. One can see the convergence behavior of the normalized wing cover mass $m_{cov,wing}$ which is shown for two different initial property sets. Here the mass values are normalized to the maximum mass value of the FEM model that is modeled with the default property set. The implemented algorithm leads in both cases to a convergence of the property values and hence to a convergence of the wing cover mass values.

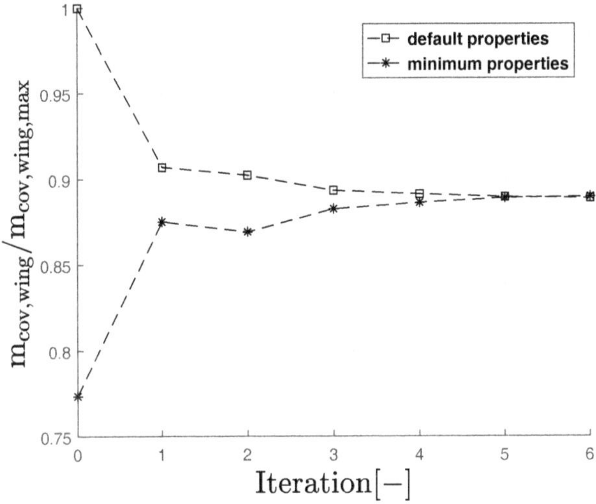

Figure 6.14: Convergence behavior of normalized wing cover mass $m_{cov,wing}$ shown for two different initial property sets

Indeed, a mass reduction is expected for the baseline model with the default initial property set since the load level along the wing is lower than the one using the complete set of sizing cases. Here the wing cover mass is reduced by a value of around 11%. Note again that the second set of initial property values is given by scaling the skin and stringer property values to predefined minimum values. The wing covers do not withstand the

applied loads with this set of minimum property values ($RF_{min} < 1.0$) and hence the property values and thus the wing cover mass are scaled up.
Further Figure 6.15 shows the evolution of the normalized wing component mass values for the same 6 iteration loops and the two different initial property sets. Here again the component mass values are normalized to the total mass of the wing optimization region in order to show the contribution of each component to the normalized total mass. It is shown that the property modification is consistent for each component and each property set, i.e. in the same iteration step both skin and stringer mass values are increased or decreased together.

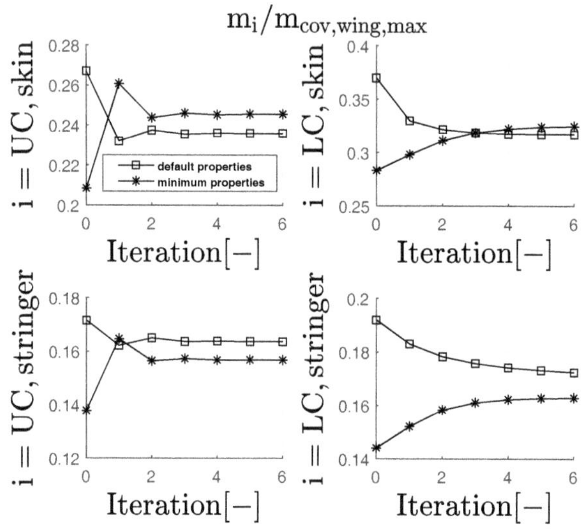

Figure 6.15: Convergence behavior of normalized wing component mass values shown for two different initial property sets

In terms of components the lower cover skin panels and stringers contribute with 37% and 19% each the highest to the wing cover total mass. Hence their mass reductions by 5% and 2% each with regard to the initial mass values that result from applying the default property set affect the evolution of the normalized wing cover total mass the most. It follows then the mass contributions of the upper cover skin panels with 27% and the upper cover stringers with 17% and their reductions by 3% and 1% respectively. Note again that in general the wing upper cover is sized by static strength criteria and the lower cover by fatigue and damage tolerance criteria. The latter ones are not covered in the implemented RF value based FSD algorithm. Hence the higher percentage mass reduction of the lower cover optimization region can be explained by this lack of failure criteria. Here the upper cover stiffened panels are dominantly sized by the flexural wrinkling criterion and the lower cover ones by the net tension criterion and in both cases with a target RF value of $RF_{min} = 1.0$. These two failure criteria are consistently dominant through all iteration loops with the different sets of property values. For this wing baseline generation the sizing load cases for both covers are dominantly vertical manoeuvres and with some exceptions vertical continuous turbulence cases with extended speedbrakes at the outer optimization region.

Another way to assess the mass changes is to calculate the concentrated mass values per slice as described earlier in chapter 5.4. Figure 6.16 shows the normalized wing concentrated mass values for two different initial property sets and the generated baseline models. Here the concentrated mass values are normalized to the total sum of wing concentrated mass values $m_{wing,conm,tot}$. Thus it can be seen how much each slice contributes to the total wing mass. The density values are assumed to be the same for all upper and lower cover stiffened panels. In addition, the swept wing whose total cross section area is getting smaller towards its tip is sliced in parallel to the streamlines. Based on these two assumptions the amplitude value of each concentrated mass depends on the number of elements that are captured inside each slice only. In this sense a smaller slice at the wing root yields a higher mass value than a bigger slice at the wing tip. Using this background information Figure 6.16 can be interpreted more meaningful. Indeed, the concentrated mass values along the normalized wing span vary for those slices that take into account the wing cover optimization region only. It is shown that the concentrated mass values converge for the two generated baseline models. On average the concentrated mass of each slice inside the optimization region is reduced by a value of 26%. This indicates that the affected skin panel and stringer properties are modified on the same scale.

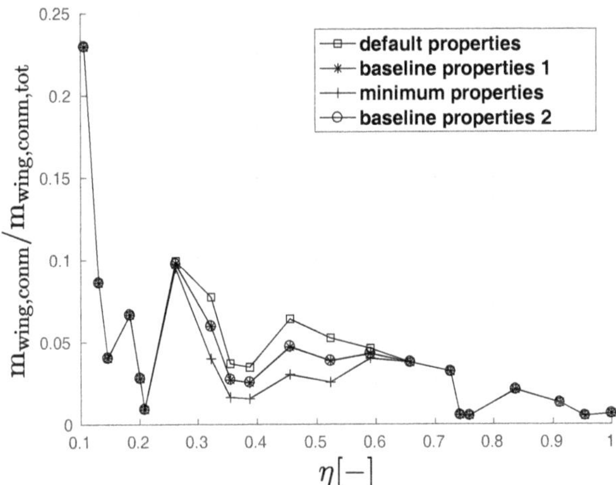

Figure 6.16: Normalized wing concentrated mass values shown for two different initial property sets and generated baseline models

In this sense the two wing cover mappings in Figure G.2 and G.3 in the appendix show the skin thickness and stringer cross section area percentage changes for the baseline generation from the default to the baseline property values. Here the upper cover skin panel thickness values are reduced by an average value of 37% and the lower cover ones by 50%. Further the upper cover stringer cross section area values are reduced by an average value of 7% and the lower cover ones by 19%. The percentage changes that are shown in the mappings confirm the results which are shown in the figures before.

The effect of altered mass distributions can also be examined by means of CG values along the wing slices. Note that the CG values here reflect the wing FEM structure only and are shown for the purpose of understanding. The concentrated mass elements and there associated CG coordinates which are used for modal analysis during the feedback loop include the fuel, system weight and other mass contributions as well. Also as mentioned in chapter 5.4 using the proposed material settings the wing structural mass accounts around 30% of the OWE only. These pieces of background information relativize the change of CG values in the scope of this structural baseline generation. In this sense Figure 6.17 shows the change of CG coordinates in x-, y- and z-direction for each slice along the normalized wing semispan. In principle CG coordinates are shifted in accordance with the related mass accumulation or removal. Here Figure 6.17 shows the normalized delta CG values between the two different initial property sets and between the two generated baseline models. Thus the convergence of both baseline models is shown.

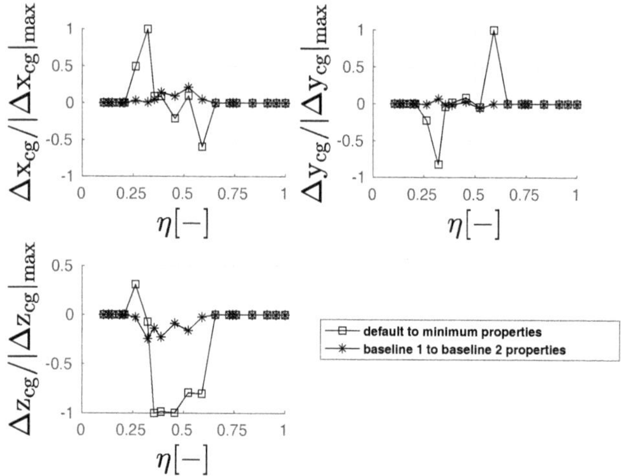

Figure 6.17: Normalized delta CG values along normalized wing semispan shown for two different initial property sets and two generated baseline models

Next to the mass changes it is also meaningful to have a look on how the updated property settings affect the wing stiffness. For this purpose the *equivalent beam* properties of the four FEM models (with default, minimum, baseline 1 and baseline 2 properties) are calculated and their delta values compared to each other for the purpose of convergence confirmation. Figure 6.18 shows the normalized delta vertical, lateral and torsional *equivalent beam* stiffness between the two different initial property sets and between the two generated baseline models.

The stiffness values behave in accordance to the mass variations. Reducing the mass results in a stiffness lessening in the very same region and vice versa. In thise sense another approach to assess the evolution of the overall stiffness during the baseline generation is to evaluate the varying maximum vertical displacement of the wing tip. Figure G.1 in the appendix shows the normalized wing tip vertical displacement U_z for the performed six iteration loops starting with the two different initial property sets and ending with the two generated baseline models. Here again the higher the mass variation the more the stiffness is affected. Together with the mass also the stiffness converges and so does the wing displacement.

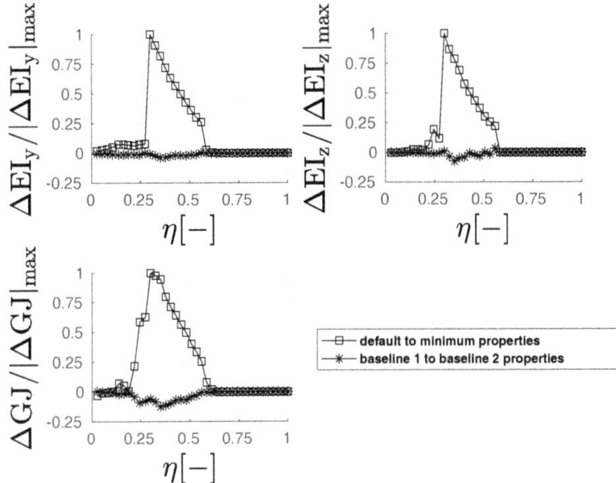

Figure 6.18: Normalized delta equivalent beam properties along normalized wing semispan shown for two different initial property sets and two generated baseline models

6.3.2 Influence of load alleviation function on structural components

In this section the impact of the load alleviation function on the wing baseline model and the aft fuselage model is shown. In the previous subsection it is shown that the RF value based FSD algorithm leads to a convergence of the wing cover properties even if different initial property sets are chosen. Hence in this subsection the *wing baseline model 1* is used only whose property set is achieved after 6 iterations by starting the FSD optimization with the default property values. In this subsection the simulations are performed with 3 different LAF parameter settings where each setting applies different control surface deflection angles and trigger threshold values for the activation of the LAF. These 3 different LAF parameter settings are explained as follows.

The *LAF parameter setting 1* applies those aileron/spoiler deflection angles and trigger threshold values as described in section 3.3.2. The implemented algorithm compares the speed V_{CAS} and vertical load factor n_z of each load case to predefined threshold values in order to decide if the LAF needs to be activated. In this sense the threshold values for the speed trigger are shown in equation 6.3 which means that the LAF is activated if the value of V_{CAS} is between $250kts$ and $340kts$.

$$250kts < V_{CAS} < 340kts \tag{6.3}$$

In addition a retraction of the outer aileron is applied according to equation 6.4. The outer aileron deflection angle is set to $\delta_{out} = 0°$ if the speed value surpasses $300kts$ accordingly.

$$300kts < V_{CAS} \tag{6.4}$$

The threshold values for the load factor n_z trigger and the related inner/outer aileron $\delta_{in,out}$ and spoiler $\delta_{SP_{4,5,6}}$ deflection angles for the vertical manoeuvre cases are shown in Figure 6.19. As it is shown on the plot at the top left the highest aileron and spoiler upwards deflections are applied when the load factor value reaches $n_z = 2.5g$. Further the highest aileron downwards deflection is applied when the load factor value reaches $n_z = -1g$.

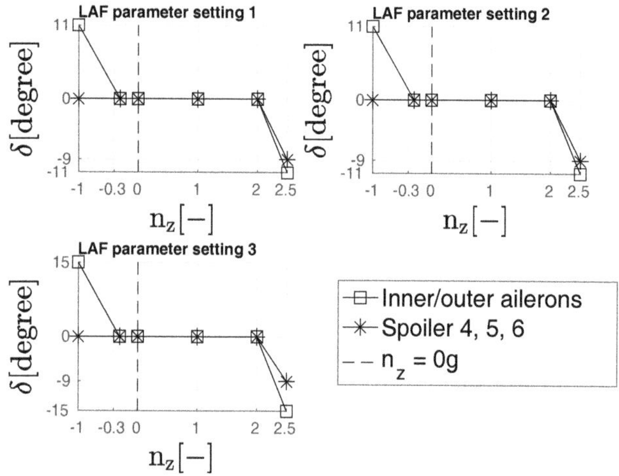

Figure 6.19: Aileron and spoiler deflection angles based on the vertical load factor shown for the LAF parameter settings 1, 2 and 3

The aileron deflection angles for the 1g LAF as well as the spoiler deflection angles for the extended speedbrakes are chosen as listed in Table 6.2. The 1g LAF is always activated if the speed trigger condition is fulfilled. Further the priority of the LAF is chosen higher than the one of the speedbrake functionality which means a retraction of the speedbrakes when the LAF is used.
The *LAF parameter setting 2* applies the same aileron and spoiler deflection angles as the first LAF parameter setting but with an altered speed trigger condition. The first LAF parameter setting with its trigger conditions does not yield a significant mass reduction for the wing component for the considered load case set as it is shown in the forthcoming figures. Indeed, when applying the first LAF parameter setting those high speed cases that are not affected by the LAF become sizing cases and thus the changes of the wing structural properties are not significant enough. Hence the outer boundary value of the speed trigger condition is increased to $370kts$ as it is shown in equation 6.5. Indeed, in

LAF parameter setting 1								
	δ_{SP_1}	δ_{SP_2}	δ_{SP_3}	δ_{SP_4}	δ_{SP_5}	δ_{SP_6}	δ_{in}	δ_{out}
1g LAF	0°	0°	0°	0°	0°	0°	-11°	-11°
Speedbrakes extended	-25°	-30°	-30°	-30°	-30°	-30°	0°	0°
LAF parameter setting 2								
	δ_{SP_1}	δ_{SP_2}	δ_{SP_3}	δ_{SP_4}	δ_{SP_5}	δ_{SP_6}	δ_{in}	δ_{out}
1g LAF	0°	0°	0°	0°	0°	0°	-11°	-11°
Speedbrakes extended	-25°	-30°	-30°	-30°	-30°	-30°	0°	0°
LAF parameter setting 3								
	δ_{SP_1}	δ_{SP_2}	δ_{SP_3}	δ_{SP_4}	δ_{SP_5}	δ_{SP_6}	δ_{in}	δ_{out}
1g LAF	0°	0°	0°	0°	0°	0°	-15°	-15°
Speedbrakes extended	-25°	-30°	-30°	-30°	-30°	-30°	0°	0°

Table 6.2: Aileron and spoiler deflection angles for 1g LAF and load cases with extended speedbrakes

Figure 6.3 a significant impact of the outer aileron on the lift distribution is shown. In this sense the outer aileron retraction for high speed cases is de-activated, too. Note that the load factor trigger stays the same as it is shown in Figure 6.19 on the plot at the top right. Furthermore the settings for the 1g LAF and the extended speedbrakes stay the same, too (cf. Table 6.2).

$$250kts < V_{CAS} < 370kts \tag{6.5}$$

The *LAF parameter setting 3* applies the same n_z and V_{CAS} trigger conditions as the second LAF parameter setting but with increased aileron deflection angles. The inner and outer aileron upwards/downwards deflection angles are increased from 11° to 15° (cf. Figure 6.19 on the plot at the bottom left and Table 6.2) in order to assess the impact of a more intense LAF on the structural components.

The impact of the different LAF parameter settings on the SMT envelope of the normalized vertical bending moment Mx along the normalized wing semispan is shown in Figure 6.20. Here the envelope values which result from the vertical manoeuvre loads calculation are shown because these cases are dominantly sizing the wing covers in this design study. Here the values of the wing baseline model are compared to those values that result when the dedicated LAF parameter setting is applied and the structural optimization performed. It can be seen that the first LAF parameter setting has no significant impact on the bending moment envelope values. The maximum positive value at the wing root is reduced by a value of 0.01% and the highest negative value by 3.5% only. The second LAF parameter setting has a high impact on the maximum positive value with a reduction of 14% whereas the reduction of the highest negative value by 3.5% equals the one with the first parameter setting. The positive and negative wing root bending moment envelope values are reduced by 17% and 4.3% when applying the third LAF parameter setting accordingly. Indeed, the applied LAF parameter settings influence the positive wing vertical bending moment the most whereas the impact on the negative values is minor. The wing upper cover is dominantly sized by compression loads and the lower cover by tension loads. A reduction of the positive wing bending moment yields smaller compression/tension loads and makes a reduction of the wing mass possible.

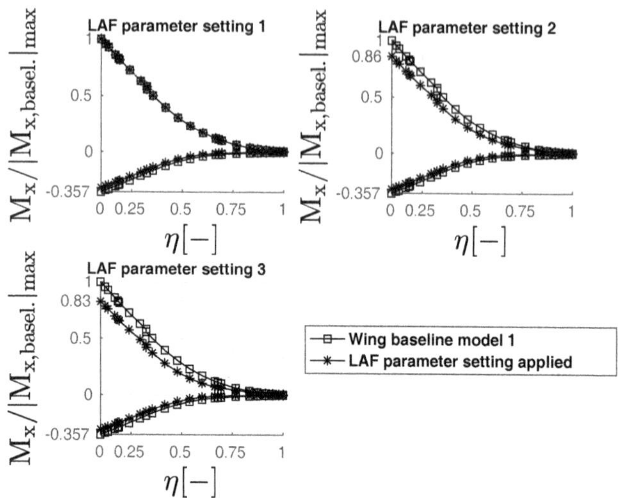

Figure 6.20: Impact of LAF parameter setting on normalized Mx SMT envelope for vertical manoeuvre cases along normalized wing semispan shown for three different LAF parameter settings

The reduction of the normalized wing cover mass $m_{cov,wing}$ based on the applied three LAF parameter settings is shown in Figure 6.21. Here the mass values are normalized to the initial wing cover total mass value of the wing baseline model. As expected, the mass reduction that is achieved with the first LAF parameter setting is neglectable. Hence the MDAO process is stopped after one iteration loop already. The second and third LAF parameter settings result in a mass reduction of 3.4% and 3.5% accordingly. The structural property values for both simulations converge after 6 iteration loops. These similar mass changes that are achieved when LAF parameter settings with different aileron deflection angles only are applied confirm the consistency of the implemented MDAO process and its algorithms.
Figure 6.22 shows the change of the normalized wing component mass values for the three different LAF parameter settings. The results for the second and third LAF parameter setting are also summarized in Table 6.3. Here again the component mass values are normalized to the initial wing cover total mass value of the wing baseline model. The achieved mass reductions when applying the first LAF parameter setting are neglectable and hence are not outlined in more detail. In terms of components the lower cover skin panels and stringers contribute with 35.7% and 19.4% each the highest to the wing cover total mass. It follows the contribution of the upper cover skin panels and stringers with 26.5% and 18.4% each. The mass reductions that are achieved using the second and third LAF parameter settings are very similar (cf. Table 6.3). The highest percentage mass reductions are achieved for those components that contribute the highest to the wing cover total mass which are the lower cover skin panels and stringers. The average percentage mass reduction of the lower cover panels is 1.55% and the one of the lower cover stringers is 0.82%. The upper cover panels and stringers are reduced by 0.68% and 0.415%.

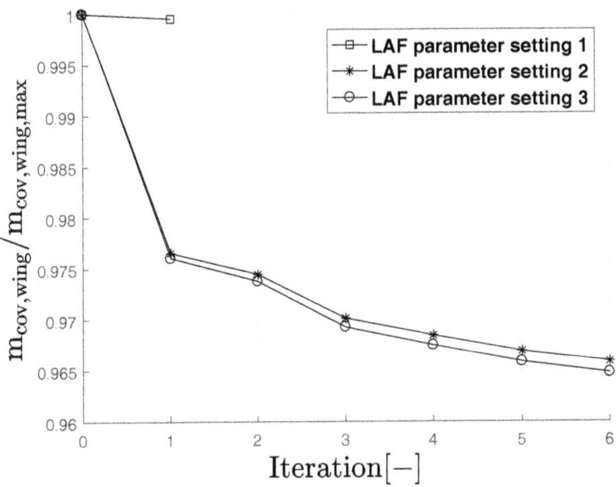

Figure 6.21: Change of normalized wing cover mass shown for three different LAF parameter settings

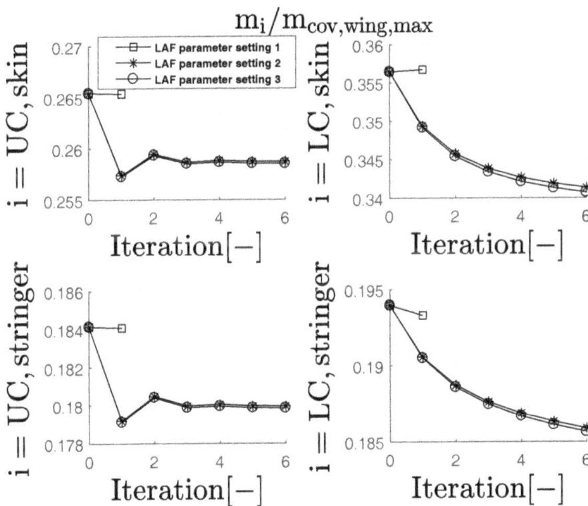

Figure 6.22: Change of normalized wing component mass values shown for three different LAF parameter settings

		$\Delta m_i/m_{cov,wing,max}$	
Component i	$m_i/m_{cov,wing,max}$	LAF parameter setting 2	LAF parameter setting 3
LC,skin	35.7%	-1.52%	-1.58%
UC,skin	26.5%	-0.67%	-0.69%
LC,stringer	19.4%	-0.81%	-0.83%
UC,stringer	18.4%	-0.41%	-0.42%

Table 6.3: Summary of normalized wing component mass reductions using the LAF parameter settings 2 and 3

One can see that the mass of the stringers and skin panels on each side of the wing box changes simultaneously. Indeed, this confirms that the FSD algorithm does not tend toward a more stringer or rather skin dominated wing cover design. This is assured by the smallest computation unit (a stiffened panel) of the implemented sizing software and also by the implemented approach of RF value based FSD property scaling. As mentioned before the highest percentage mass reduction is achieved for the lower cover skin panels, followed by the percentage reduction for the upper cover ones. In the ranking the stringers follow accordingly. Note again that one reason for this high percentage mass reduction for the lower cover is the lack of fatigue and damage tolerance sizing criteria.
The percentage changes for each skin panel and stringer inside the optimization region of both covers and for the two LAF parameter settings are shown in the mappings in the Figures H.2, H.3, H.4 and H.5 in the appendix. Here the average (of all considered skin panels of the dedicated wing cover) percentage skin thickness reductions of the upper cover for the second and third LAF parameter settings are 10.93% and 11.25%. The average skin thickness reduction values for the lower cover are 25.81% and 26.31%. The average stringer cross section area reduction values for the upper cover are 7.70% and 7.91% and for the lower cover 13.52% and 13.79% accordingly. Also here it can be seen that the reduction values for the two LAF parameter settings are very similar to each other.
For both LAF parameter settings the dominant sizing cases with a target RF value of $RF_{min} = 1.0$ are vertical manoeuvre cases with retracted and extended speedbrakes. Here again the upper cover is dominantly sized by the flexural wrinkling criterion and the lower cover by the net tension criterion.
Further the normalized concentrated mass values along the normalized wing semispan for the baseline model and the optimized wing component using the two different LAF parameter settings are shown in Figure 6.23. Beside the wing mappings, which make a post processing across ribs and stringers possible, this kind of curves allow the assessment of mass changes per slice in a quick way. Here again each concentrated mass value is normalized to the total sum of concentrated mass values along the wing semispan in order to see how much each concentrated mass contributes to the total mass. The concentrated mass values of those slices that take into account the optimization region are reduced by an average value of 9.86%. The resulting concentrated mass values confirm the conclusions that are drawn from the previously shown figures and tables.
The property changes at the upper and lower covers have a minor impact on the CG values of the concentrated masses along the wing. In the frame of the performed property changes the wing box structure is considered only without taking into account components like the slat, flap, system equipment or the fuel tanks. Based on this and the arguments regarding the composition of the wing mass which are mentioned in section 5.4 the dimension of

the here calculated change of CG values is even more irrelevant. However the developed software offers the possibility to quantify and plot the changes of CG values along the x-axis (trailing to leading edge), y-axis (root to tip) and the z-axis (lower to upper cover).

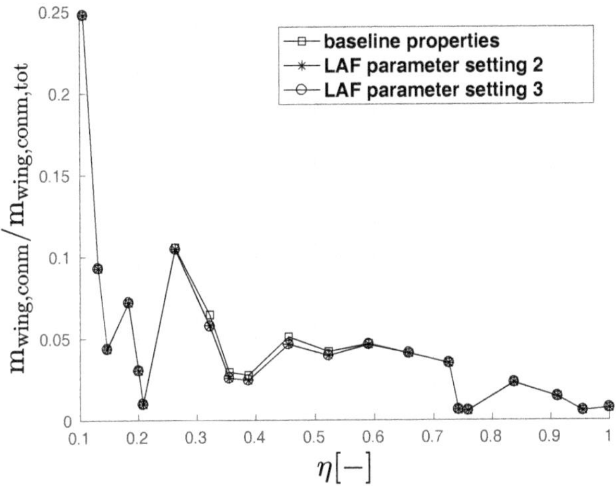

Figure 6.23: Normalized concentrated mass values along normalized wing semispan shown for baseline properties and two different LAF parameter settings

In this sense the change of the normalized CG values along the normalized wing semispan is shown for the two different LAF parameter settings in Figure 6.24. Here the CG values that are calculated after the structural property optimization are normalized to the CG values of the baseline model. Thus it is possible to assess the changes of the newly calculated CG values with regard to the x_{cg}, y_{cg} and z_{cg} values of the baseline model. The changes of the CG values are very similar when using the two different LAF parameter settings. Further the changes of the x_{cg} and y_{cg} are with values less than 1% neglectable. The z_{cg} values are affected the most by the structural property changes of the wing covers. Here the highest decrease amounts to 23.39% and the highest increase is 11.1%. However also the changes of the z_{cg} values are in the scale of millimeters and hence not significant. The effect of skin thickness and stringer cross section area changes on the wing stiffness is assessed with the *equivalent beam* method. Figure 6.25 shows the change of the normalized *equivalent beam* stiffness values EI_y, EI_z and GJ along the normalized wing semispan for the two different LAF parameter settings. The stiffness changes for both LAF parameter settings are characteristically the same. It can be seen that the property changes affect the three stiffness quantities by percentage in the same way and are proportional to the related property modifications. In this sense a higher reduction of skin thickness and stringer cross section area means a higher stiffness decrease. Further the evolution of the maximum wing tip deflection in vertical direction is another means of analyzing the impact on the wing stiffness. According to this Figure H.1 in the appendix shows the change of the normalized wing tip vertical displacement U_z for the two different LAF parameter settings. One can see how the maximum wing tip deflection increases by 3.2% with regard to the decreasing stiffness.

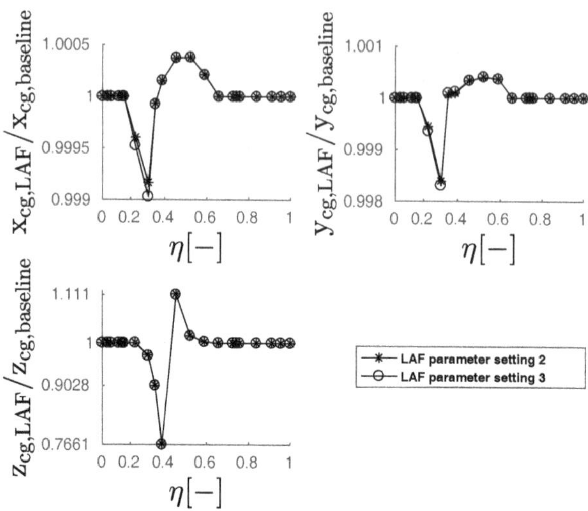

Figure 6.24: Change of normalized CG values along normalized wing semispan shown for two different LAF parameter settings

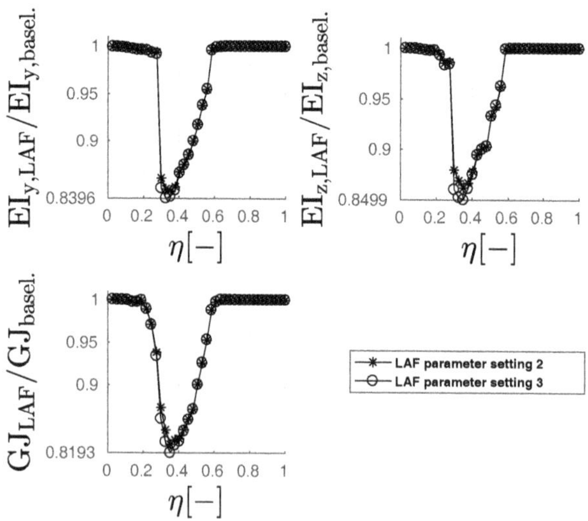

Figure 6.25: Change of normalized equivalent beam properties along normalized wing semispan shown for two different LAF parameter settings

Since the achieved mass reductions using the two different LAF parameter settings are nearly similar the question arises if it brings more added value to set the aileron and spoiler deflection angles beyond a certain threshold value for the purpose of load alleviation. As described before in section 3.3.2 high control surface deflection angles can yield disadvantages in the frame of handling qualities. Note again that handling qualities concern the requirements regarding the manoeuvrability of an aircraft by the pilot during climb, cruise, landing and further conditions. In this sense for instance it is shown in Figure 6.5 in section 6.1 that for a 2.5g manoeuvre even when applying both ailerons the shift of the center of lift converges to a value of about 16% for aileron deflection angles which are lower than -10°. The same convergence behaviour is shown for the wing root bending moment in Figure 6.6. With respect to handling quality targets (cf. section 3.3.2) and the here calculated mass reductions of the wing covers using the two different LAF parameter settings this kind of argumentation can be used before considering higher or lower aileron deflection angles for the purpose of load alleviation and structural optimization.

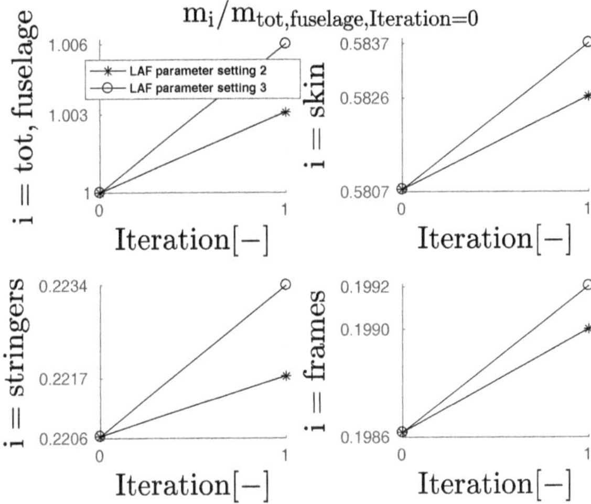

Figure 6.26: Fuselage mass penalty estimation with 536 load cases - changes of normalized component mass values of the analyzed aft fuselage region shown for two LAF parameter settings

The resulting mass penalty of the aft fuselage section is estimated by the previously described internal loads exceedance calculation. As explained before for the here performed analysis no stress threshold value is used and hence the complete set of FEs that are included in the analysis region are considered. The SMT loads that result along the fuselage are recalculated using the wing model properties after the final iteration loop of each LAF analysis scenario. The new set of fuselage global FEM internal loads is compared to the baseline values by applying the predefined exceedance criteria. Figure 6.26 shows the normalized mass increases for both LAF parameter settings after one iteration loop of the internal loads exceedance method. It can be seen that for both LAF parameter settings the estimated total mass increase (plot at the top left) is below 1%. It is shown

that the skin panels contribute the highest to the total mass of the analyzed aft fuselage region, followed by the stringers and the frames.
Indeed, the estimated mass penalties are neglectable and furthermore for future investigations an alternative and more sophisticated approach needs to be investigated. Comparable to the wing box optimization approach the fuselage elements need to undergo a more detailed local strength analysis based on various sizing criteria. The estimated mass penalty is then the summation of local mass increases due to the property changes.
Another important aspect for the structural assessment due to the LAF parameter changes is the location of the aircraft CG for each mass case. In general the location of the CG along the fuselage varies with the positioning of cabin and cargo items and the available fuel. Indeed, an optimal range for the location of the CG exists within the allowable limits which are shown in the weight and balance diagram. In this sense the aircraft CG location and the related total weight of the chosen five mass cases for the purpose of design studies is shown in Figure 3.2 in section 3.1. One can see the quantitative CG envelope for a generic long range aircraft and the weight values for MZFW, MLW and MTOW. The studies show that the investigated force increase at the aft fuselage region due to the pitch compensation appears for those mass cases whose CG are located more aft (behind 40%) and whose mass values are close to MLW.

7 Summary, discussion of methods and results, outlook

7.1 Summary

The major objectives that are set for the making of this PhD thesis are accomplished. In this sense dedicated in-house Airbus tools and self-developed algorithms are implemented into a multidisciplinary analysis and optimization framework as a whole. The framework can be used to produce reliable and accurate results computing hundreds of altering load conditions and applying external discrete forces and moments to the global finite element models for the purpose of structural property optimization and mass penalty estimation in an appropriate time frame. Multiple iterations of the whole process chain can be performed with updates of the mass and stiffness distribution of the loads calculation model. The physics of each module is meaningful and correctly represented which thus allows to perform parameter studies for a better judgment of an activated load alleviation system on mass variations of the wing and fuselage. This achievement is one step forward to support or rather replace today's means of scaling data bases or pure expert opinion. In chapter 2 an overview of the developed process chain and its computational implementation into the Airbus infrastructure is given.
Chapter 3 describes the loads calculation process in more detail. The scope of computed load cases are vertical balanced manoeuvres and continuous turbulence cases for different flight points and mass cases. No time simulations are performed but rather correlated load cases are computed instead with respect to the subsequent structural sizing. The loads calculation model consists in particular of a beam stick model, a lumped mass model, combined with an aerodynamic panel method and a simulation model of the *electronic flight control system*. The load alleviation function parameter variation is implemented in form of an interface that steers the aileron and spoiler deflection angles. The results of the loads calculation process are quantified by SMT values which are also used as the major indicator for the impact of the load alleviation function.
Also general aspects of the *electronic flight control system* are described with a special attention on the load alleviation function which is used to alleviate the loads along the wing. In this sense an activated load alleviation function shifts the center of pressure inboard and thus leads to a reduction of the wing root bending moment and torque. The side effect of an activated load alleviation function is the related pitch compensation by an elevator demand which yields an additional load increase at the aft fuselage section. Currently the load alleviation function parameter variation is performed in form of a sensitivity study and not in an optimization loop. Indeed, the variation of the aileron and outer spoiler deflection angles for vertical conditions in the frame of a sensitivity study is a proper means of showing the impact on the SMT values of the correlated load cases. The fact that correlated load cases are needed for the structural sizing makes this aspect of the loads calculation process sufficient. However for better understanding a complete optimization of the flight control system and in particular its load alleviation function-

ality has to be tackled using time simulations including a parametrization of gains, time constants and considering aspects of flight and handling quality.
Different trigger threshold values are applied for the activation of the load alleviation function and to consider sizing relevant load cases inside a predefined range of speed and load factor values. In order to perform the design studies with a set of flight conditions that follows certification specifications one has to take into account cases which are not affected by the load alleviation function, too. Especially these cases can advance to sizing relevancy when the original critical cases are alleviated. The same aspect applies to the diversification of load conditions by the center of gravity and mass cases according to the weight and balance diagram. The most important mass cases with respect to the additional load increase at the aft fuselage section due to the pitch compensation are those whose center of gravity is located more aft.
In chapter 4 the structural property optimization and mass penalty estimation approaches are explained. For this purpose the material properties and modeling aspects of the global finite element models are explained. It follows a description of the external loading of both the fuselage and wing component. It is common practice to apply the external forces to the wing global finite element model along the wing reference axis. Still one can think of more sophisticated methods like nodal loads in which the discrete forces and moments are applied across the wing covers which makes the load introduction more realistic. The fuselage loading is based on the *unitary load case* approach which is more accurate than the load distribution using rigid body elements although the set of *unitary load cases* needs to be reassessed for each new aircraft.
The altered fully stressed design algorithm based on stress values shows a fast convergence behavior and is well suited to gain a better understanding of this way of structural property modification in a reasonable time frame. The altered fully stressed design algorithm is based on a single failure criterion. However the wing needs to be sized with more sophisticated failure criteria. On the other hand the reserve factor based fully stressed design approach is based on the minimum reserve factor value of multiple failure criteria which is used to update the stiffened panel structural properties. Accordingly the reserve factor based fully stressed design approach yields more accurate results. Indeed, the necessary sizing process takes a couple of hours in contrast to the purely stress based approach that runs a couple of minutes only. A common side effect of the fully stressed design approach is the checkerboard pattern of the optimized region. This effect can be overcome by suppression algorithms and with additional predefined sizing regions for the purpose of harmonization and to comply with manufacturing constraints. In order to go a step further mathematical optimization using gradient based approaches are the only reasonable alternative for so many design variables and today's computing resources to get results in a proper time frame.
The wing structural property optimization takes a couple of hours for one complete analysis loop. Hence the idea is to use a less time consuming approach for the fuselage mass penalty estimation which is realized by internal loads exceedance envelopes. These envelopes are built with 1-dimensional criteria and 2-dimensional convex hulls, too. Note that the implementation of more sophisticated failure criteria for another component next to the wing requires significantly more development time. This way of structural assessment based on the internal loads of a global finite element model is less time consuming but strongly depends on the quality of both the baseline and newly calculated envelope. In the scope of this thesis static sizing criteria are implemented only which is in especially sufficient for regions like the upper wing cover. Structural fatigue life analysis and residual

strength analysis of damaged structures are not investigated. Indeed, the lack of these criteria affects the results of the structural property optimization since regions like the lower wing cover and parts of the fuselage stiffened shell are predominantly sized by these criteria. In addition to the drawback of missing sizing criteria also the applied set of load cases is incomplete. In this sense significant ground conditions are excluded completely. This is one major reason why the regions around the landing gear attachment points are excluded from the optimization since the property values there are very high and so are the possible mass reductions without the original sizing cases.

In chapter 5 both the *equivalent beam* approach and the grid point weight generator are explained which are the two main implementations into the framework in order to send back updated structural properties to the loads calculation model in form of delta stiffness and mass values. The *equivalent beam* approach which is used to pass the change of wing stiffness from one global finite element model to another is sufficient enough in the scope of this thesis. Alternatively the condensed stiffness matrix itself can be used and thus couple terms in the off-diagonal entries are included in the subsequent analysis. However the discrepancies in the discretization of both global finite element models are too significant. On the other hand stiffness contributions of components that are not explicitly modeled in the finite element model are also included with a chain of beam elements in today's way of working in industry. The use of the grid point weight generator for the calculation of concentrated mass values is also sufficient enough since changes of the wing primary structure are assessed only. The delta of concentrated mass values is set as the global convergence criterion which is appropriate while the use of delta stiffness values is not mature enough in the current status of this process.

In chapter 6 the performed design studies are shown. The current version of the developed framework deals with active load alleviation which is a use case for vertical manoeuvres including 1g cases only. Hence landing approaches and conditions on the ground or lateral cases are neither investigated nor included in this thesis. This leads to the matter that the loads level of the applied cases is not the same as the one that is achieved by the original set of sizing cases. Hence the idea of baseline load cases and a related baseline model of the wing box is chosen. In this sense for the baseline generation the load alleviation function is deactivated for all load conditions and the wing box covers are sized with the reserve factor based fully stressed design approach taking into account the implemented failure criteria and running the whole analysis loop until mass convergence. In accordance to this the internal loads envelopes are generated for the different parts of the aft fuselage section using the same set of baseline load cases. Here the considered use case is the analysis of already sized structural models after preliminary design where an incremental update of the flight control is incorporated. Hence one can categorize this use case as a niche one. Indeed, a change of the structural wing box properties to a new baseline after this state of maturity is an unconventional approach but still an appropriate way of working when the target is to assess the delta of load alleviation function and mass relations.

For the finally performed design studies the load alleviation function is applied with different parameter settings and the whole process chain is run until mass convergence of the wing box is achieved. For the applied load alleviation function parameter settings the analysis of the wing box converges after six iterations at most with a maximum mass reduction of 3.5%. The highest portion of this reduction is accounted to the property changes of the lower cover skin panels and stringers which are indeed predominantly sized by fatigue and damage tolerance criteria. Hence another option for a more conservative assessment of the mass reduction can be to exclude the complete lower cover from the optimization. Here the approach of a previously prepared baseline is chosen but as another solution one can exclude all those structural regions that are not sized by the considered set of load conditions and focus only on small fragments of the wing covers instead. But then again it is still needed to take into account the possible change of critical load case ranking due to the effect of load alleviation. The activated load alleviation function leads to a reduction of both SMT values at the wing root and structural mass at the optimization region of the wing covers. The internal loads exceedance analysis is performed for all finite elements of the considered aft fuselage section and all stress levels. The mass penalty estimation shows no significant impact of the activated load alleviation function and the related pitch compensation on the aft fuselage region. Indeed, the way of mass penalty estimation can be enhanced with more sophisticated approaches like static plus fatigue and damage tolerance criteria.

7.2 Discussion of methods

The developed simulation framework is applicable but indeed the results and their interpretation depend on the chosen circumstances (load cases, failure criteria, allowable values, optimized regions, etc.) for this kind of niche use case. Here the structural assessment of load alleviation function parameter variations for an already sized aircraft model is performed. Due to the lack of the complete set of sizing load cases and failure criteria a baseline model of the wing is generated first which considers dedicated vertical balanced manoeuvres, vertical continuous turbulence cases and static strength failure criteria only. This baseline model of the wing enables a relative assessment of wing structural mass reductions and fuselage mass penalties due to load alleviation function parameter variations. The computation takes a couple of hours for the chosen amount of load cases. Despite this, the amount of information that one gets out of the simulation process is worth the computational effort. Note that otherwise in the current industrial aircraft development process those pieces of information, that can be retrieved from the developed simulation framework, need to be requested explicitly from the responsible department. The processing of such an inquiry can take weeks or even longer.
The computation of hundreds of load conditions based on the 1-dimensional loads stick model and aerodynamic panel model is very robust and time efficient. Together with the aerodynamic databases for correction this loads calculation process yields sufficiently accurate aircraft reactive loads in order to assess the influence of load alleviation function parameter variations properly. In this sense the developed routines for the quantification and visualization of the aircraft reactive loads and other interesting quantities like the lift distribution are a proper means to speed up the post processing of data after the loads calculation process. Additionally, the developed interface to set and vary parameters of the *electronic flight control system* for the purpose of load alleviation enables a quick modification of dedicated control surface deflection angles for hundreds of load conditions with considered speed and vertical load factor trigger threshold values.
The structural optimization of the wing component by a reserve factor based fully stressed design scaling algorithm is the most time consuming part of the whole simulation framework. Indeed, the use of parallel computing enables the reduction of computation time in relation to the number of cores that is used very well. On the other hand alternative approaches like gradient based optimization algorithms that rely on the calculation

of sensitivities are even more time consuming. Nevertheless, the use of gradient based optimization techniques promises more accurate results and hence is an important topic for future studies. The predominant sizing of the upper covers by the flexural wrinkling criterion and of the lower covers by the net tension criterion shows that the distribution of tension/compression loads among the wing covers is appropriately realized. In addition, the implemented optimization algorithm leads to a convergence of the structural mass. However the implemented fully stressed design based optimization is partially validated and the achieved mass reductions are too optimistic. In future studies it is needed to consider more sophisticated failure criteria, like fatigue and damage tolerance criteria, and more sizing relevant load conditions, too.
The estimation of mass penalties based on internal loads exceedance values is neither time consuming nor computationally expensive. However, the implemented approach does not yield a convergence of mass values. Additionally, inconsistencies like the calculation of high exceedance values for reduced stress values and the low stress level of the analyzed aft fuselage region based on the applied load case set make this approach look inefficient. Hence in future studies more sophisticated means like failure envelopes or structural sizing based on fatigue and damage tolerance criteria need to be investigated.
The implemented *equivalent beam* approach and grid point weight generator that are used in order to send back delta stiffness and delta mass values to the loads calculation model are computationally inexpensive and fulfill their purposes appropriately. However the harmonization of the different finite element models in order to be able to work with condensed stiffness and mass matrices can lead to results with higher accuracy in future studies.
The combination of self-developed program code and algorithms together with Airbus in-house tools and commercial software into a single simulation framework as a whole is achieved. The results and generated data sets of the simulation runs are repeatable and physically meaningful. The framework is developed in an industrial context and is directly placed on the Airbus computing infrastructure. The framework is built as generic as possible in order to enable the analysis and optimization of other aircraft simulation models, too. Even though some restrictions are not avoidable. Especially the input data that is needed for the structural sizing software, which is an Airbus in-house development, has to be prepared according to Airbus specifications. Similar restrictions apply for the preparation of data that is needed for the factorization process of the *unitary load cases*. In order to enable the use of the developed simulation framework in an industrial context special attention is paid to user friendliness. In this sense next to a structured source code, which is also commented, a steering file in ASCII text format is provided that contains the required file paths and switch parameters. Thus other engineers can use the simulation framework as a black box tool.

7.3 Discussion of results

The activated load alleviation function leads to a reduction of the wing root bending moment and torque and thus enables a mass reduction of the wing upper and lower covers. The developed algorithms and routines make a reliable quantification possible. Due to the reduced set of sizing load cases and failure criteria a relative assessment is possible only. Hence the analysis and optimization region needs to be selected with engineering judgment according to the considered load cases and failure criteria. Further the selection of load cases that are affected by the load alleviation function depends on the predefined speed and load factor trigger threshold values. On the other hand those load cases that are not affected can become sizing relevant and impact the outcome of the wing structural property optimization. A set of recommended trigger threshold values is used for a first design study. These threshold values yield a neglectable impact of the load alleviation function due to those load cases that are not affected and hence become sizing relevant. In future studies it is needed to further adapt the trigger threshold values or consider more sizing relevant load cases which are affected by the load alleviation function, too. The estimated mass penalties for the aft fuselage shell are neglectable. In future studies it is needed to apply more sophisticated approaches for the analysis of the aft fuselage section. One possibility is the use of sizing relevant failure criteria which can make the estimation of mass penalties more accurate.

7.4 Outlook

In future studies one can think of different ways to improve the overall multidisciplinary simulation framework. First of all the harmonization of both the global finite element *model for strength assessment* and the *model for dynamic analysis* can enable the exchange of the stiffness properties with condensed matrices instead of using *equivalent beam* properties. In a second step the implementation of more static, fatigue and damage tolerance sizing criteria for both components together with mathematical optimization and the definition of sizing regions can lead to an improved structural assessment. In terms of flight control dedicated time simulations can lead to a better integration of this discipline into the whole multidisciplinary framework. In terms of computing resources if one thinks a couple of years forward then the use of *quantum computing* can have a deep impact on the complete way of working. The finite element simulation models can be adapted to 3-dimensional elements, CFD calculations can be used more extensively in a much reduced time frame, a full aircraft model can be used over the whole process chain and big data can be transformed into artificial intelligence much faster.

Bibliography

[ACY+10] J. Ainsworth, C. Collier, P. Yarrington, R. Lucking, and J. Locke. Airframe wingbox preliminary design and weight prediction. In: *69th Annual Conference on Mass Properties, Virginia Beach, Virginia*. Society of Allied Weight Engineers, 2010.

[Air] Airbus. A330 - Flight crew operating manual.

[Air00] Airbus. A340 Flight Deck and Systems Briefing for Pilots. URL: `https://www.smartcockpit.com/docs/A340_Flight_Deck_and_Systems_Briefing_For_Pilots.pdf`, 2000.

[Air19] Airbus. Global Market Forecast - Cities, Airports & Aircraft. URL: `https://www.airbus.com/content/dam/corporate-topics/strategy/global-market-forecast/GMF-2019-2038-Airbus-Commercial-Aircraft-book.pdf`, 2019.

[Akm06] A. Akmeşe. *Aeroservoelastic analysis and robust controller synthesis for flutter suppression of air vehicle control actuation systems*. PhD thesis, Department of Mechanical Engineering, Middle East Technical University, 2006.

[AMB03] C. Anhalt, H. P. Monner, and E. Breitbach. Interdisciplinary Wing Design - Structural Aspects. In: *World Aviation Congress 2003*, 2003.

[And02] B. Andoleit. *Eine Parallelisierungsstrategie zur numerischen Strömungssimulation für Flüssigkeiten mit Gedächtnis*. PhD thesis, Fakultät Maschinenbau, Professur für Strömungsmechanik, Helmut-Schmidt-Universität, 2002.

[AR69] E. Albano and W. P. Rodden. Doublet-Lattice Method for Calculating Lift Distributions on Oscillating Surfaces in Subsonic Flows. *AIAA Journal*, 7:279–285, 1969.

[Ard06] M. D. Ardema. *Newton-Euler Dynamics*. Springer Science & Business Media, 2006.

[Arg54] J. Argyris. Flexure-Torsion Failure of Panels: A Study of Instability and Failure of Stiffened Panels under Compression when Buckling in Long Wavelengths. *Aircraft Engineering and Aerospace Technology*, 26(6):174–184, 1954.

[BAG97] J. M. Biannic, P. Apkarian, and W. L. Garrard. Parameter Varying Control of a High Performance Aircraft. *Journal of Guidance, Control, and Dynamics*, 20:225–231, 1997.

[BAL11] R. Brockhaus, W. Alles, and R. Luckner. *Flugregelung*. Springer, 2011.

[BB06] C. Boller and M. Buderath. Fatigue in aerostructures - where structural health monitoring can contribute to a complex subject. *Philosophical Transactions of the Royal Society A: Mathematical, Physical and Engineering Sciences*, 365(1851):561–587, 2006.

[BCDR+08] A. Bérard, L. Cavagna, A. Da Ronch, L. Riccobene, S. Ricci, and A. T. Isikveren. Development and validation of a next-generation conceptual aero-structural sizing suite. In: *ICAS 2008, 26th International Congress of the Aeronautical Sciences*, 2008.

[BGW99] M. L. Baker, P. J. Goggin, and B. A. Winther. Aeroservoelastic modeling, analysis, and design techniques for transport aircraft. Technical report ADP010476, Boeing Phantom Works, 1999.

[BJSH66] M. R. Barber, C. K. Jones, T. R. Sisk, and F. W. Haise. An evaluation of the handling qualities of seven general-aviation aircraft. Technical report NASA TN D-3726, National Aeronautics and Space Administration, 1966.

[Bla92] M. Blair. A compilation of the mathematics leading to the doublet lattice method. Technical report WL-TR-92-3028, Wright Laboratory, 1992.

[Bon09] C. Bonnet. Optimum CG position - What is the best CG position for an aircraft? In: *16th Performance and Operations conference*, 2009.

[BSB00] J. Brink-Spalink and J. Bruns. Correction of unsteady aerodynamic influence coefficients using experimental or CFD data. In: *41st Structures, Structural Dynamics, and Materials Conference and Exhibit*, 2000.

[BW90] P. Bartholomew and H. K. Wellen. Computer-aided optimization of aircraft structures. *Journal of Aircraft*, 27(12):1079–1086, 1990.

[BZA87] C. S. Buttrill, T. A. Zeiler, and P. D. Arbuckle. Nonlinear simulation of a flexible aircraft in maneuvering flight. In: *Flight Simulation Technologies Conference*, 1987.

[Cir11] R. Cirillo. Detailed and condensed finite element models for dynamic analysis of a business jet aircraft. Master's thesis, School of Industrial and Information Engineering, Politecnico Milano, 2011.

[CRR09] L. Cavagna, S. Ricci, and L. Riccobene. A fast tool for structural sizing, aeroelastic analysis and optimization in aircraft conceptual design. In: *50th AIAA/ASME/ASCE/AHS/ASC Structures, Structural Dynamics, and Materials Conference*, 2009.

[Deb01] K. Deb. *Multi-Objective Optimization using Evolutionary Algorithms*. John Wiley & Sons, Ltd, 2001.

[Dha15] A. Dharmasaroja. *Efficient modelling of failure envelopes and load patterns in aircraft structures.* PhD thesis, School of Mechanical and Aerospace Engineering, Queen's University Belfast, 2015.

[DKAG13] J. K. S. Dillinger, T. Klimmek, M. M. Abdalla, and Z. Gürdal. Stiffness optimization of composite wings with aeroelastic constraints. *Journal of Aircraft*, 50(4):1159–1168, 2013.

[DNG12] F. Dorbath, B. Nagel, and V. Gollnick. A knowledge based approach for extended physics-based wing mass estimation in early design stages. In: *28th International Congress of the Aeronautical Sciences*, 2012.

[dWASS+07] O. de Weck, J. Agte, J. Sobieszczanski-Sobieski, P. Arendsen, A. Morris, and M. Spieck. State-of-the-art and future trends in multidisciplinary design optimization. In: *48th AIAA/ASME/ASCE/AHS/ASC Structures, Structural Dynamics, and Materials Conference*, 2007.

[ESA09] M. S. A. Elsayed, R. Sedaghati, and M. Abdo. Accurate Stick Model Development for Static Analysis of Complex Aircraft Wing-Box Structures. *AIAA Journal*, 47:2063–2075, 2009.

[FLW02] R. Faye, R. Laprete, and M. Winter. Blended winglets for improved airplane performance. Technical report AERO QTR 01 2002, The Boeing Company, 2002.

[Fuc16] M. B. Fuchs. *Structures and their analysis.* Springer, 2016.

[FWFH14] T. Führer, C. Willberg, S. Freund, and F. Heinecke. Parametric model generation and automated sizing process for the analysis of aircraft structures using the design environment DELiS. In: *4th European Aeronautics Science Network International Workshop*, 2014.

[Gar18] E. Garrigues. A Review of Industrial Aeroelasticity Practices at Dassault Aviation for Military Aircraft and Business Jets. *AerospaceLab Journal*, (14):1–34, 2018.

[GB84] H. G. Giesseler and G. Beuck. Design procedure of an active load alleviation system. In: *ICAS - The International Council of the Aeronautical Sciences*, 1984.

[GCdCV08] P. Gasbarri, L. D. Chiwiacowsky, and H. F. de Campos Velho. A hybrid multilevel approach for aeroelastic optimization of composite wing-box. In: *EngOpt 2008 - International Conference on Engineering Optimization*, 2008.

[Ger14] M. Gerhardt. Untersuchung des Einflusses von Temperatur- und Innendrucklasten im Entwurf eines modernen Transportflugzeuges mithilfe von Ersatzmodellen. Master's thesis, Institut für Flugzeugbau und Leichtbau, Technische Universität Braunschweig, 2014.

[Giu97] A. A. Giunta. *Aircraft multidisciplinary design optimization using design of experiments theory and response surface modeling methods.* PhD thesis, Faculty of Virginia Polytechnic Institute and State University, Virginia Tech, 1997.

[GKB09] S. Grihon, L. Krog, and D. Bassir. Numerical optimization applied to structure sizing at AIRBUS: a multi-step process. *International Journal for Simulation and Multidisciplinary Design Optimization*, 3(4):432–442, 2009.

[Gri17] S. Grihon. Structure sizing optimization capabilities at AIRBUS. In: *World Congress of Structural and Multidisciplinary Optimisation*, pages 719–737. Springer, 2017.

[GSM+09] S. Grihon, M. Samuelides, A. Merval, A. Remouchamps, M. Bruyneel, B. Colson, and K. Hertel. Fuselage Structure Optimisation. *Advances in Collaborative Civil Aeronautical Multidisciplinary Design Optimization*, 2009.

[Guy65] R. J. Guyan. Reduction of Stiffness and Mass Matrices. *AIAA Journal*, 3:380–380, 1965.

[Hag12] S. Haghighat. *Multidisciplinary Design Optimization of A Highly Flexible Aeroservoelastic Wing.* PhD thesis, Institute for Aerospace Studies, University of Toronto, 2012.

[Han09] L. U. Hansen. *Optimierung von Strukturbauweisen im Gesamtentwurf von Blended Wing Body Flugzeugen.* PhD thesis, Institut für Flugzeugbau und Leichtbau, Technische Universität Braunschweig, 2009.

[Hed66] S. G. Hedman. Vortex lattice method for calculation of quasi steady state loadings on thin elastic wings in subsonic flow. Technical report FFA Report 105, FFA - The Aeronautical Research Institute of Sweden, 1966.

[HHH08] L. U. Hansen, W. Heinze, and P. Horst. Blended wing body structures in multidisciplinary pre-design. *Structural and Multidisciplinary Optimization*, 36(1):93–106, 2008.

[HML12] S. Haghighat, J. R. R. A. Martins, and H. H. T. Liu. Aeroservoelastic design optimization of a flexible wing. *Journal of Aircraft*, 49(2):432–443, 2012.

[HP11] D. H. Hodges and G. A. Pierce. *Introduction to Structural Dynamics and Aeroelasticity.* Cambridge University Press, 2011.

[Hür10] F. Hürlimann. *Mass Estimation of Transport Aircraft Wingbox Structures with a CAD/CAE-Based Multidisciplinary Process.* PhD thesis, Centre of Structure Technologies, ETH Zurich, 2010.

[Inc17] Alcoa Mill Products Inc. Alcoa - Alloy 2024 - Sheet and Plate. 2017.

[Jam12] K. James. *Aerostructural shape and topology optimization of aircraft wings*. PhD thesis, Graduate Department of Institute for Aerospace Studies, University of Toronto, 2012.

[JBL+02] H. D. Joos, J. Bals, G. Looye, K. Schnepper, and A. Varga. A multi-objective optimisation-based software environment for control systems design. In: *IEEE International Symposium on Computer Aided Control System Design Proceedings*, 2002.

[Joh02] E. H. Johnson. Fully Stressed Design in MSC. Nastran. In: *3rd MSC. Software Worldwide Aerospace Users Conference and Technology Showcase*, 2002.

[Joo96] H. D. Joos. Multi-Objective Parameter Synthesis - MOPS. Technical report GARTEUR / TP-088-16, Group for Aeronautical Research and Technology in Europe - GARTEUR, 1996.

[Joo99] H. D. Joos. A methodology for multi-objective design assessment and flight control synthesis tuning. Technical report TR R323-99, Deutsches Institut für Luft- und Raumfahrt - Institut für Robotik und Systemdynamik, 1999.

[JPM10] P. W. Jansen, R. E. Perez, and J. R. R. A. Martins. Aerostructural Optimization of Nonplanar Lifting Surfaces. *Journal of Aircraft*, 47:1490–1503, 2010.

[JV88] E. H. Johnson and V. B. Venkayya. Automated Structural Optimization System (ASTROS): Volume 1 - Theoretical Manual. Technical report, Air Force Wright Aeronautical Laboratories, 1988.

[KDDB+16] W. R. Krüger, J. Dillinger, R. De Breuker, M. Reyes, and K. Haydn. Adaptive Wing: Investigations of Passive Wing Technologies for Loads Reduction in the CleanSky Smart Fixed Wing Aircraft (SFWA) Project. In: *Greener Aviation 2016*, 2016.

[KDDBH19] W. R. Krüger, J. Dillinger, R. De Breuker, and K. Haydn. Investigations of passive wing technologies for load reduction. *CEAS Aeronautical Journal*, 10(4):977–993, 2019.

[KFN17] C. Kaiser, D. Friedewald, and J. Nitzsche. Comparison of nonlinear CFD with time-linearized CFD and CFD-corrected DLM for gust encounter simulations. In: *International Forum on Aeroelasticity and Structural Dynamics - IFASD 2017*, 2017.

[KG97] R. Kelm and M. Grabietz. Aeroelastic effects on the weight of an aircraft in the pre-design phase. In: *57th Annual Conference of the Society of Allied Weight Engineers*. Society of Allied Weight Engineers, 1997.

[KG00] R. Kelm and M. Grabietz. Method of reducing wind gust loads acting on an aircraft. Patent US006161801A, 2000. URL: `https://patentimages.storage.googleapis.com/41/14/76/15d03df964da0f/US6161801.pdf`.

[KK16] W. R. Krüger and T. Klimmek. Definition of a comprehensive loads process in the DLR project iLOADS. In: *Deutscher Luft- und Raumfahrtkongress 2016*, 2016.

[KKM12] G. K. W. Kenway, G. J. Kennedy, and J. R. R. A. Martins. A scalable parallel approach for high-fidelity aerostructural analysis and optimization. In: *53rd AIAA/ASME/ASCE/AHS/ASC Structures, Structural Dynamics, and Materials Conference*, 2012.

[KLG95] R. Kelm, M. Läpple, and M. Grabietz. Wing primary structure weight estimation of transport aircraft in the pre-development phase. In: *54th Annual Conference of Society of Allied Weight Engineers*, 1995.

[KM12] G. J. Kennedy and J. R. R. A. Martins. A comparison of metallic and composite aircraft wings using aerostructural design optimization. In: *14th AIAA/ISSMO Multidisciplinary Analysis and Optimization Conference*, 2012.

[KM14] G. K. W. Kenway and J. R. R. A. Martins. Aerostructural optimization of the Common Research Model configuration. In: *15th AIAA/ISSMO Multidisciplinary Analysis and Optimization Conference*, 2014.

[KOCH16] T. Klimmek, P. Ohme, P. D. Ciampa, and V. Handojo. Aircraft Loads - An Important Task from Pre-Design to Loads Flight Testing. In: *Deutscher Luft- und Raumfahrtkongress 2016*, 2016.

[KP91] J. Katz and A. Plotkin. *Low-speed aerodynamics: From wing theory to panel methods*. McGraw-Hill, 1991.

[KPL52] P. Kuhn, J. P. Peterson, and L. R. Levin. A summary of diagonal tension - part 1: methods of analysis. Technical report NACA-TN-2661, National Advisory Committee for Aeronautics, 1952.

[Krü08] W. R. Krüger. A multi-body approach for modelling manoeuvring aeroelastic aircraft during preliminary design. *Proceedings of the Institution of Mechanical Engineers, Part G: Journal of Aerospace Engineering*, 222(6):887–894, 2008.

[KV06] E. Kesseler and W. J. Vankan. Multidisciplinary design analysis and multi-objective optimisation applied to aircraft wing. *Life*, 65(85):95–101, 2006.

[Lew89] G. E. Lewis. Maneuver load alleviation system. Patent US4796192, 1989. URL: `https://patentimages.storage.googleapis.com/06/eb/45/507067228b8122/US4796192.pdf`.

[LHH15] K. Lindhorst, M. Haupt, and P. Horst. Aeroelastic analyses of the high-Reynolds-number-aerostructural-dynamics configuration using a nonlinear surrogate model approach. *AIAA Journal*, 53(9):2784–2796, 2015.

[LJ01] G. Looye and H. D. Joos. Design of robust dynamic inversion control laws using multi-objective optimization. In: *AIAA Guidance, Navigation, and Control Conference and Exhibit*, 2001.

[LL00] D. J. Leith and W. E. Leithead. Survey of Gain-Scheduling Analysis and Design. *International Journal of Control*, 73:1001–1025, 2000.

[Lom96] T. L. Lomax. *Structural Loads Analysis for Commercial Transport Aircraft: Theory and Practice.* American Institute of Aeronautics and Astronautics, 1996.

[LSX01] Q. Li, G. P. Steven, and Y. M. Xie. A simple checkerboard suppression algorithm for evolutionary structural optimization. *Structural and Multidisciplinary Optimization*, 22(3):230–239, 2001.

[Lub94] A. A. Lubis. *Zur Optimierung des multidisziplinären Entwurfs von Verkehrsflugzeugen in der parallelen Rechenumgebung.* PhD thesis, Institut für Flugzeugbau und Leichtbau, Technische Universität Braunschweig, 1994.

[M+81] R. MacNeal et al. *The Nastran theoretical manual.* National Aeronautics and Space Administration, 1981.

[Mar78] G. L. Martin. Paneling techniques for use with the Vorlax computer program. Technical report NAS1-13500, Vought Corporation Hampton Technical Center, 1978.

[MAR01] J. R. R. A. Martins, J. J. Alonso, and J. Reuther. Aero-structural wing design optimization using high-fidelity sensitivity analysis. In: *Proceedings - CEAS Conference on Multidisciplinary Aircraft Design Optimization*, 2001.

[MC02] J. H. McMasters and R. M. Cummings. Airplane Design - Past, Present, and Future. *Journal of Aircraft*, 39(1):10–17, 2002.

[Meg10] T. H. G. Megson. *An Introduction to Aircraft Structural Analysis.* Butterworth-Heinemann, 2010.

[Mel00] T. Melin. A Vortex Lattice MATLAB Implementation for Linear Aerodynamic Wing Applications. Master's thesis, Department of Aeronautics - Royal Institute of Technology (KTH), 2000.

[Mes15] A. Messac. *Optimization in Practice with Matlab - For Engineering Students and Professionals.* Cambridge University Press, 2015.

[Miu90] H. Miura. An improved fully stressed design algorithm for plate/shell structures. *Structural optimization*, 2(4):233–237, 1990.

[ML13] J. R. R. A. Martins and A. B. Lambe. Multidisciplinary Design Optimization: A Survey of Architectures. *AIAA Journal*, 51(9):2049–2075, 2013.

[MTS82] G. J. McRae, J. W. Tilden, and J. H. Seinfeld. Global sensitivity analysis - a computational implementation of the Fourier Amplitude Sensitivity Test (FAST). *Computers & Chemical Engineering*, 6(1):15–25, 1982.

[Niu99] M. C. Y. Niu. *Airframe stress analysis and sizing*. Conmilit Press Ltd., 1999.

[NLHS15a] R. Nazzeri, F. Lange, M. Haupt, and C. Sebastien. Assessing sensitivities of maneuver load alleviation parameters on buckling reserve factors using surrogate model based extended Fourier amplitude sensitivity test. In: *11th World Congress on Structural and Multidisciplinary Optimisation*, 2015.

[NLHS15b] R. Nazzeri, F. Lange, M. Haupt, and C. Sebastien. Selection of critical load cases using an artificial neural network approach for reserve factor estimation. In: *Deutscher Luft- und Raumfahrtkongress 2015*. Deutsche Gesellschaft für Luft- und Raumfahrt - Lilienthal-Oberth e.V., 2015.

[NRW93] R. A. Nichols, R. T. Reichert, and J. R. Wilson. Gain Scheduling for H-Infinity Controllers: A Flight Control Example. *IEEE Transactions on Control Systems Technology*, 1:69–79, 1993.

[OH17] M. Opie and D. Hadcroft. C-27J NASTRAN Global Finite Element Model User Manual C27J-DST-v3.0. Technical report DST-Group-TN-1700, Australian Government, Department of Defence: Aerospace Division - Defence Science and Technology Group, 2017. URL: https://www.dst.defence.gov.au/sites/default/files/publications/documents/DST-Group-TN-1700.pdf.

[ÖHH00] C. M. Österheld, W. Heinze, and P. Horst. Influence of aeroelastic effects on preliminary aircraft design. In: *Proceedings of the 22nd International Congress of the Aeronautical Sciences*, 2000.

[ÖHH01] C. M. Österheld, W. Heinze, and P. Horst. Preliminary design of a blended wing body configuration using the design tool PrADO. In: *CEAS conference on multidisciplinary aircraft design and optimisation*, 2001.

[Öst06] C. M. Österheld. *Physikalisch begründete Analyseverfahren im integrierten multidisziplinären Flugzeugvorentwurf*. PhD thesis, Institut für Flugzeugbau und Leichtbau, Technische Universität Braunschweig, 2006.

[PAK04] P. Piperni, M. Abdo, and F. Kafyeke. The application of multi-disciplinary optimization technologies to the design of a business jet. In: *10th AIAA/ISSMO Multidisciplinary Analysis and Optimization Conference*, 2004.

[Pal11] N. Paletta. *Maneuver load controls, analysis and design for flexible aircraft*. PhD thesis, PhD school of Aerospace, Naval and Quality Engineering, University of Napoli, 2011.

[Pet16] F. B. Peters. Analyse des Einflusses von lokal skalierten Schub-Moment-Torsion Kurven auf die Flugzeugstruktur mit Hilfe von Sensitivitätsanalysemethoden unter Verwendung von Ersatzmodellen. Master's thesis, Institut für Flugzeugbau und Leichtbau, Technische Universität Braunschweig, 2016.

[PGB93] S. N. Patnaik, J. D. Guptill, and L. Berke. Merits and Limitations of Optimality Criteria Method for Structural Optimization. Technical report NASA Technical Paper 3373, National Aeronautics and Space Administration, 1993.

[PH98] S. N. Patnaik and D. A. Hopkins. Optimality of a fully stressed design. Technical report NASA TM-1998-207411, National Aeronautics and Space Administration, 1998.

[PHEL09] H. Panzer, J. Hubele, R. Eid, and B. Lohmann. Generating a Parametric Finite Element Model of a 3D Cantilever Timoshenko Beam Using Matlab. Technical report, Institute of Automatic Control, Technical University of Munich, 2009.

[PJB+13] K. H. Park, S. O. Jun, S. M. Baek, M. H. Cho, K. J. Yee, and D. H. Lee. Reduced-order model with an artificial neural network for aerostructural design optimization. *Journal of Aircraft*, 50(4):1106–1116, 2013.

[PLB06] R. E. Perez, H. H. T. Liu, and K. Behdinan. Multidisciplinary optimization framework for control-configuration integration in aircraft conceptual design. *Journal of Aircraft*, 43(6):1937–1948, 2006.

[POK+19] M. Pusch, D. Ossmann, T. Kier, M. Tang, J. Lübker, and J. Dillinger. Aeroelastic Modeling and Control of an Experimental Flexible Wing. In: *AIAA Scitech 2019 Forum*, 2019.

[Raz65] R. Razani. Behavior of fully stressed design of structures and its relationship to minimum-weight design. *AIAA Journal*, 3(12):2262–2268, 1965.

[Res06] C. Reschke. *Integrated Flight Loads Modelling and Analysis for Flexible Transport Aircraft*. PhD thesis, Institut für Flugmechanik und Flugregelung der Universität Stuttgart, 2006.

[Rie13] J. Rieke. *Bewertung von CFK Strukturen in einem multidisziplinären Entwurfsansatz für Verkehrsflugzeuge*. PhD thesis, Institut für Flugzeugbau und Leichtbau, Technische Universität Braunschweig, 2013.

[RJ04] W. P. Rodden and E. H. Johnson. *MSC. Nastran - Aeroelastic Analysis - User's Guide*. MSC. Software, 2004.

[RK07] W. Rüther-Kindel. *Beiträge zur flugmechanischen Modellierung in zeitvariablen, räumlichen Windfeldern für die Echtzeit-Flugsimulation*. PhD thesis, Fakultät für Verkehrs- und Maschinensysteme, Technische Universität Berlin, 2007.

[RQ14] B. Raju and D. Quinn. Rapid Sizing of a Composite Wing Structure. In: *4th Aircraft Structural Design Conference*. Royal Aeronautical Society, 2014.

[Rum12] M. P. Rumpfkeil. Optimization Under Uncertainty Using Derivatives and Kriging Surrogate Models. In: *Seventh International Conference on Computational Fluid Dynamics - ICCFD7*, 2012.

[RZ91] G. I. N. Rozvany and M. Zhou. A note on truss design for stress and displacement constraints by optimality criteria methods. *Structural Optimization*, 3:45–50, 1991.

[Sad12] M. Sadraey. *Aircraft Design: A Systems Engineering Approach*. Wiley Publications, 2012.

[SGN10] E. L. Salih, L. Gardner, and D. A. Nethercot. Numerical investigation of net section failure in stainless steel bolted connections. *Journal of Constructional Steel Research*, 66(12):1455–1466, 2010.

[SJR78] L. A. Schmit Jr. and R. K. Ramanathan. Multilevel Approach to Minimum Weight Design including Buckling Constraints. *AIAA Journal*, 16(2):97–104, 1978.

[SKDS13] J. Scherer, D. Kohlgrüber, F. Dorbath, and M. Sorour. A finite element based tool chain for structural sizing of transport aircraft in preliminary aircraft design. In: *Deutscher Luft- und Raumfahrtkongress 2013*, 2013.

[SML+16] K. Sommerwerk, B. Michels, K. Lindhorst, M. Haupt, and P. Horst. Application of efficient surrogate modeling to aeroelastic analyses of an aircraft wing. *Aerospace Science and Technology*, 55:314–323, 2016.

[SP56] J. W. Semonian and J. P. Peterson. An analysis of the stability and ultimate compressive strength of short sheet-stringer panels with special reference to the influence of the riveted connection between sheet and stringer. Technical report NACA-TN-1255, National Advisory Committee for Aeronautics, 1956.

[SS82] J. Sobieszczanski-Sobieski. A linear decomposition method for large optimization problems - blueprint for development. Technical report NASA Technical Paper 83248, National Aeronautics and Space Administration, 1982.

[SS88] J. Sobieszczanski-Sobieski. Optimization by decomposition - a step from hierarchic to non-hierarchic systems. Technical report NASA Technical Paper 101494, National Aeronautics and Space Administration, 1988.

[SSBA83] J. Sobieszczanski-Sobieski, J. Benjamin, and D. Augustine. Structural optimization by multilevel decomposition. Technical report NASA Technical Paper 84641, National Aeronautics and Space Administration, 1983.

[SSF+16] A. Schuster, J. Scherer, T. Führer, T. Bach, and D. Kohlgrüber. Automated sizing process of a complete aircraft structure for the usage within a MDO process. In: *Deutscher Luft- und Raumfahrtkongress 2016*. Deutsche Gesellschaft für Luft- und Raumfahrt - Lilienthal-Oberth e.V., 2016.

[STC99] A. Saltelli, S. Tarantola, and K. P.-S. Chan. A Quantitative Model-Independent Method for Global Sensitivity Analysis of Model Output. *Technometrics*, 41(1):39–56, 1999.

[Ste12] P. Steinke. *Finite-Elemente-Methode - Rechnergestützte Einführung*. Springer, 2012.

[Sun98] C. T. Sun. *Mechanics of aircraft structures*. John Wiley & Sons, Inc., 1998.

[SY93] S. Suzuki and S. Yonezawa. Simultaneous structure/control design optimization of a wing structure with a gust load alleviation system. *Journal of Aircraft*, 30(2):268–274, 1993.

[TD14] R. Thormann and D. Dimitrov. Correction of aerodynamic influence matrices for transonic flow. *CEAS Aeronautical Journal*, 5:435–446, 2014.

[USA86] USAF. Flying qualities textbook - Volume 2, Part 1. Technical report USAF-TPS-CUR-86-02, USAF Test Pilot School, 1986.

[VCO17] A. Voss, G. P. Chiozzotto, and P. Ohme. Dynamic maneuver loads calculation for a sailplane and comparison with flight test. In: *International Forum on Aeroelasticity and Structural Dynamics - IFASD 2017*, 2017.

[VdVKKM00] A. Van der Velden, D. Kokan, R. Kelm, and J. Mertens. Application of MDO to large subsonic transport aircraft. In: *38th Aerospace Sciences Meeting and Exhibit*, 2000.

[Ven89] V. B. Venkayya. Optimality criteria: a basis for multidisciplinary design optimization. *Computational Mechanics*, 5:1–21, 1989.

[VK17] A. Voss and T. Klimmek. Design and sizing of a parametric structural model for a UCAV configuration for loads and aeroelastic analysis. *CEAS Aeronautical Journal*, 8(1):67–77, 2017.

[VSS04] G. Venter and J. Sobieszczanski-Sobieski. Multidisciplinary optimization of a transport aircraft wing using particle swarm optimization. *Structural and Multidisciplinary Optimization*, 26(1):121–131, 2004.

[WC10] L. Wiart and G. Carrier. Accounting for Wing Flexibility in the Aerodynamic Calculation of Transport Aircraft Using Equivalent Beam Model. In: *13th AIAA/ISSMO Multidisciplinary Analysis Optimization Conference*, 2010.

[WC15] J. R. Wright and J. E. Cooper. *Introduction to Aircraft Aeroelasticity and Loads*. John Wiley & Sons, Ltd, 2015.

[WD90] G. A. Wrenn and A. R. Dovi. Multilevel Decomposition Approach to the Preliminary Sizing of a Transport Aircraft Wing. Technical report NASA Technical Paper 4296, National Aeronautics and Space Administration, 1990.

[WKP+18] M. Wüstenhagen, T. Kier, M. Pusch, D. Ossmann, Y. M. Meddaikar, and A. Hermanutz. Aeroservoelastic Modeling and Analysis of a Highly Flexible Flutter Demonstrator. In: *2018 Atmospheric Flight Mechanics Conference, AIAA AVIATION Forum*, 2018.

[WSG12] J. Wenzel, M. Sinapius, and U. Gabbert. Primary structure mass estimation in early phases of aircraft development using the finite element method. *CEAS Aeronautical Journal*, 3:35–44, 2012.

[WSTA+17] W. Weigold, B. Stickan, I. Travieso-Alvarez, C. Kaiser, and P. Teufel. Linearized Unsteady CFD for Gust Loads with TAU – IFASD 2017. In: *International Forum on Aeroelasticity and Structural Dynamics - IFASD 2017*, 2017.

[XK11a] J. Xu and I. Kroo. Aircraft Design with Maneuver and Gust Load Alleviation. In: *29th AIAA Applied Aerodynamics Conference*, 2011.

[XK11b] J. Xu and I. Kroo. Aircraft Design with Maneuver Load Alleviation and Natural Laminar Flow. In: *11th AIAA Aviation Technology, Integration, and Operations (ATIO) Conference*, 2011.

[Xu12] J. Xu. *Aircraft design with active load alleviation and natural laminar flow.* PhD thesis, Department of aeronautics and astronautics, Stanford University, 2012.

[YM12] W. Yamazaki and D. J. Mavriplis. Derivative-enhanced variable fidelity surrogate modeling for aerodynamic functions. *AIAA Journal*, 51(1):126–137, 2012.

[ZD84] B. Ziegler and M. Durandeau. Flight control system on modern civil aircraft. In: *ICAS - The International Council of the Aeronautical Sciences*, 1984.

[Zot92] R. Zotemantel. MBB-LAGRANGE: A general structural reliability and optimization structural system. In: *Reliability and Optimization of Structural Systems' 91*, pages 427–442. Springer, 1992.

[ZY07] W. Zhang and Z. Ye. Control law design for transonic aeroservoelasticity. *Aerospace Science and Technology*, 11(2-3):136–145, 2007.

A Loads calculation and structural assessment

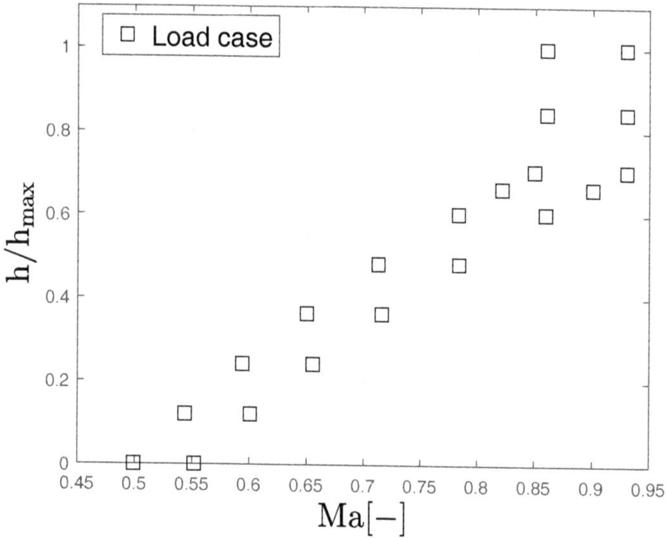

Figure A.1: Mach - altitude diagram of applied load cases for design studies

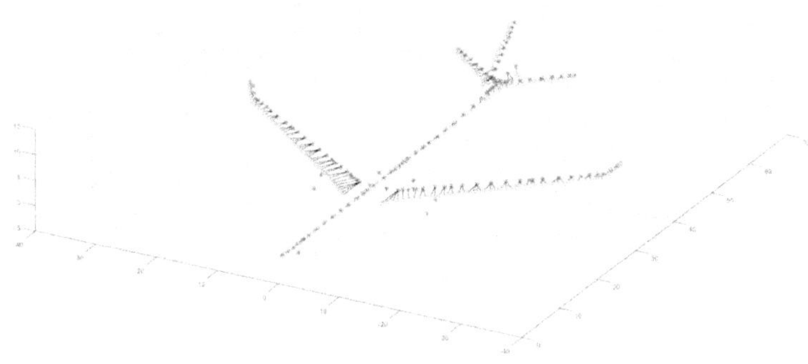

Figure A.2: Generic long range aircraft aeroelastic model: structural grids + splining

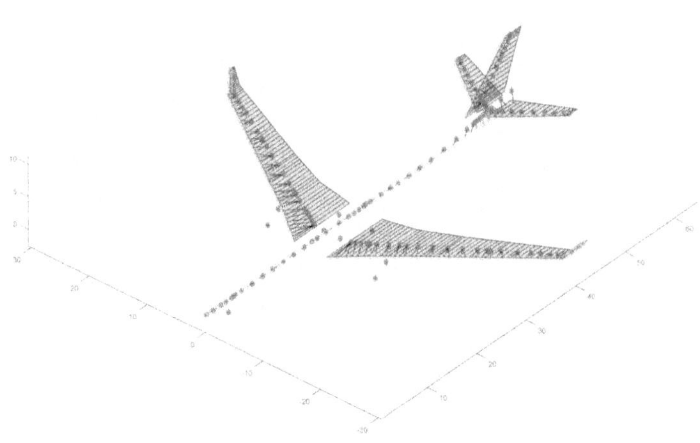

Figure A.3: Generic long range aircraft aeroelastic model: structural grids + aero panels + splining

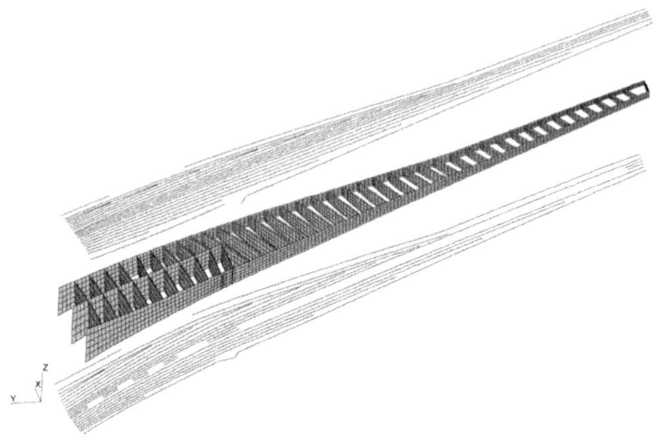

Figure A.4: Global FEM of generic aircraft wing box model - exploded view of skin cover stringers

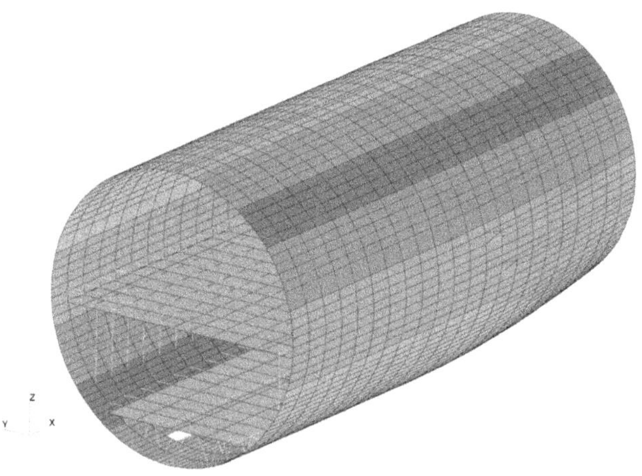

Figure A.5: Global FEM of generic aircraft fuselage model

Aluminum alloy 2024 - T4	
Property	**Value**
E	73,000 MPa
G	28,000 MPa
ν	0.3
ρ	2.78 g/cm^3
$R_{p0.2}$	276 MPa
R_m	427 MPa

Table A.1: Aluminum alloy material properties for global FEM

B Feedback loop - main process steps

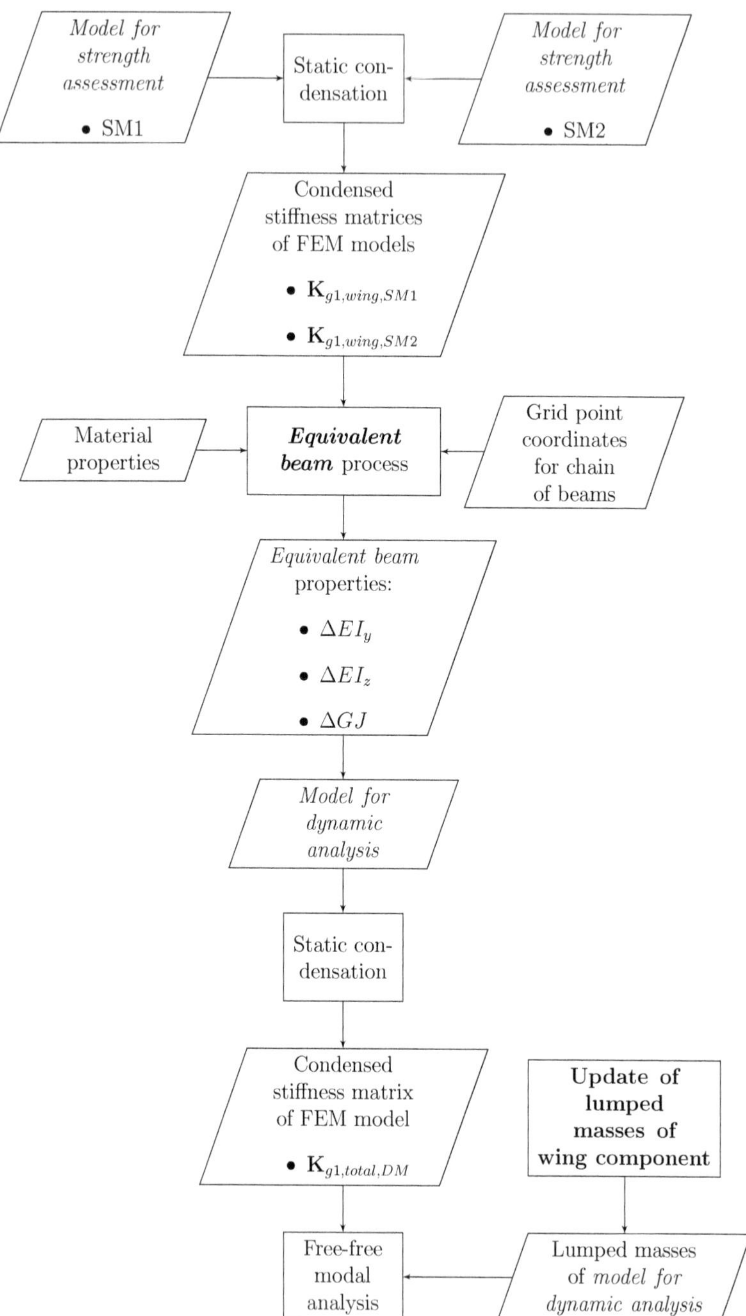

Figure B.1: Overall process steps - feedback loop - part 1

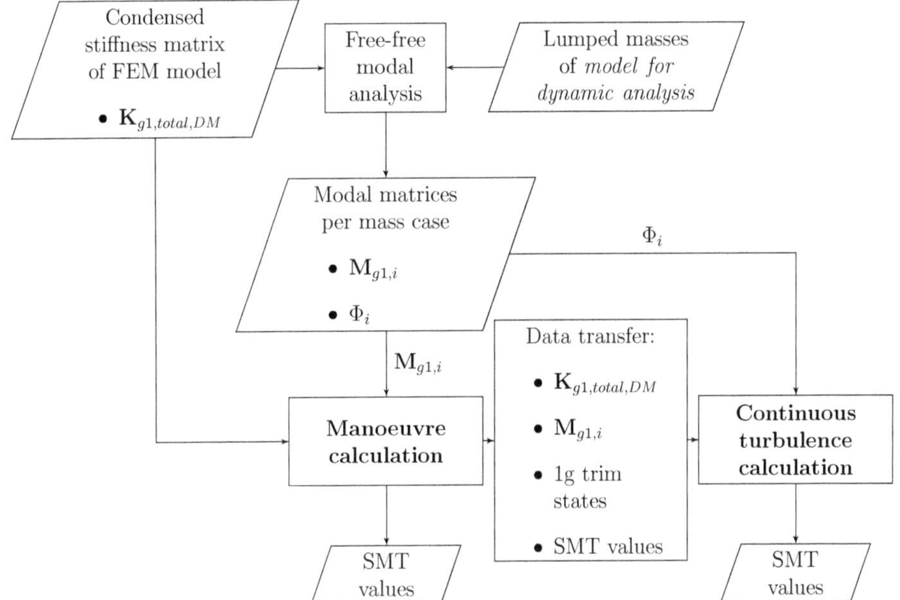

Figure B.2: Overall process steps - feedback loop - part 2

C Equivalent beam approach - validation

Vertical balanced manoeuvre	
Quantity	**Value**
n_z	$2.5g$
Ma	0.7
h	$10007ft$
V_{CAS}	$365kts$

Table C.1: Characteristic values of validation case

In order to validate the *equivalent beam* approach one of the vertical balanced manoeuvre cases, which are also used for the design studies, is applied to the different forms of the wing structure. Here the wing FEM *model for strength assessment*, the condensed form of this FEM according to Guyan and the representation by the chain of beams with its *equivalent beam* properties are used. These 3 models are loaded with the same manoeuvre case for the purpose of comparing their resulting deflections to each other. Some characteristic values of the test load condition are given in Table C.1. Despite other validation approaches which use unit loads at the wing tip only a real simulated load condition is chosen here to make sure that relatively high deflection values are compared to each other.
Only the resulting values for the translational degree of freedom u_z are shown here in Figure C.1 because the highest deflection values are reached in z-direction whereas the movements in the other two directions are comparably small. The wing coordinate system is the same as shown in Figure 5.1. Note that the FEM model of the wing is clamped at the center wing box which means that the degrees of freedom of the first grid point that is evaluated are not restricted. But still the displacement values for this grid point are close to zero since its location is close to the center wing box.

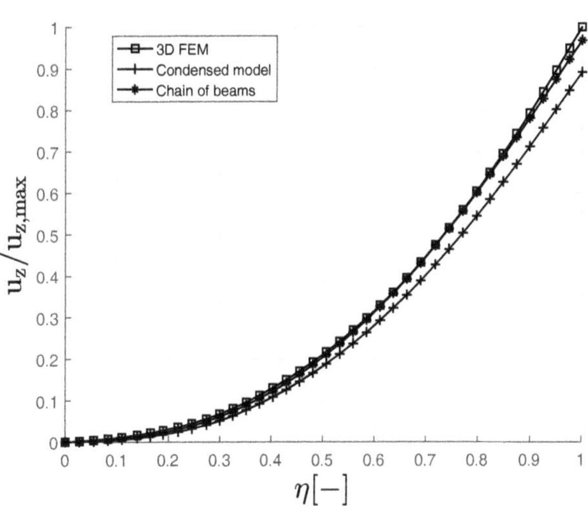

Figure C.1: Comparison of degree of freedom: u_z

D Equivalent beam approach - influence of skin and stringer

In order to have a better understanding of the influence of upper and lower wing cover properties on the resulting *equivalent beam* stiffness values a small variation analysis is performed. Here the skin panel thickness values and stringer cross sectional parameters of the very first initial wing FEM *model for strength assessment* (starting point for the baseline generation) are downscaled by 50%. Each time one quantity is factorized and its *equivalent beam* stiffness values are compared with the initial ones. The results can be seen in the following Figures D.1, D.2 and D.3. One can conclude that the skin panel thickness values influence the *equivalent beam* stiffness more than the stringer parameters. In the same way the upper cover properties influence the *equivalent beam* stiffness more than the lower cover ones.

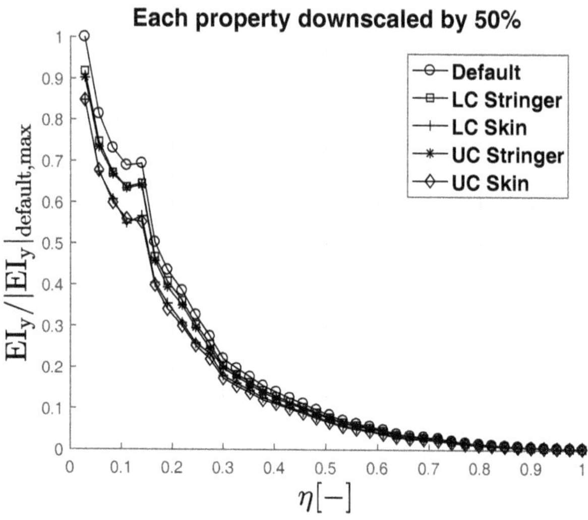

Figure D.1: Influence of property scaling on vertical equivalent beam stiffness

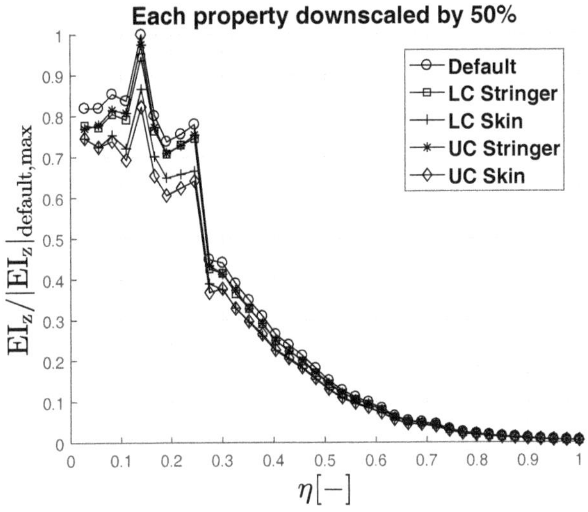

Figure D.2: Influence of property scaling on lateral equivalent beam stiffness

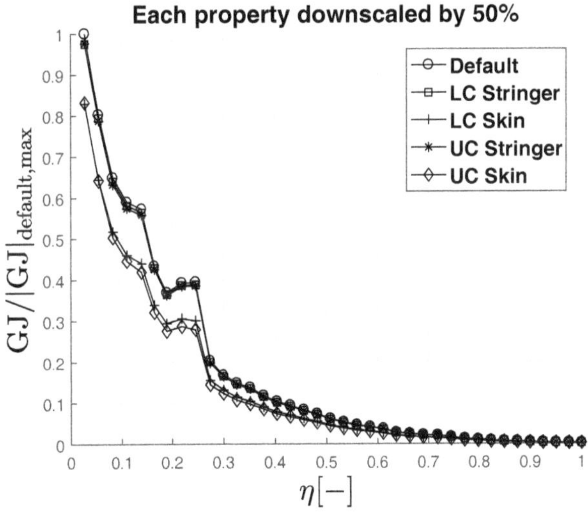

Figure D.3: Influence of property scaling on torsional equivalent beam stiffness

E Grid Point Weight Generator - basics

In this chapter the basic equations behind the MSC Patran® GPWG according to [M+81] are presented. When applying the GPWG to a FEM model it will output its density based mass, its main and secondary inertias as well as its CG coordinates with respect to a reference point. As described in chapter 5.4 in the frame of this research the inertias are scaled according to the mass changes and the CG values are assumed to stay the same. Hence the only information that is needed from the GPWG is the mass per FE of the targeted elements in the model. In this sense the main necessary steps to calculate these mass values are described only.
In a first step the GPWG calculates the rigid body transformation matrix $\mathbf{G}$ based on acceleration vectors for the whole FE model. As shown in equation E.1 this matrix transforms the grid point acceleration values $\ddot{\mathbf{u}}_0$ which are in the local coordinate system to values $\ddot{\mathbf{u}}_{gg}$ in the global coordinate system.

$$\ddot{\mathbf{u}}_{gg} = \mathbf{G}\, \ddot{\mathbf{u}}_0 \tag{E.1}$$

In a next step the rigid body transformation matrix is applied to the initial mass matrix $\mathbf{M}_{gg}$ (cf. equation E.2) which is given in global coordinates and whose size is not reduced yet due to single- or multi-point constraints. The resulting mass matrix $\mathbf{M}_0$ contains the rigid body mass properties with its values given in the local coordinate system. The transformation itself can be performed using any reference point but in general the origin of the local coordinate system is chosen.

$$\mathbf{M}_0 = \mathbf{G}^T\, \mathbf{M}_{gg}\, \mathbf{G} \tag{E.2}$$

Equation E.3 shows the rigid body mass matrix $\mathbf{M}_0$ which is of the size 6×6 and consists of a translational part $\mathbf{M}_t$, a rotational part $\mathbf{M}_r$ and further combined parts $\mathbf{M}_{tr}$ and $\mathbf{M}_{rt}$.

$$\mathbf{M}_0 = \begin{bmatrix} \mathbf{M}_t & \mathbf{M}_{tr} \\ \mathbf{M}_{rt} & \mathbf{M}_r \end{bmatrix} \tag{E.3}$$

The density based mass values of the targeted elements are retrieved from the translational part $\mathbf{M}_t$ inside the rigid body mass matrix. Equation E.4 shows how this diagonal matrix is defined.

$$\mathbf{M}_t = \begin{bmatrix} \mathbf{M}_{t,11} & 0 & 0 \\ 0 & \mathbf{M}_{t,22} & 0 \\ 0 & 0 & \mathbf{M}_{t,33} \end{bmatrix} \tag{E.4}$$

In general the mass values $\mathbf{M}_{t,11}$, $\mathbf{M}_{t,22}$ and $\mathbf{M}_{t,33}$ are the same for each of the three coordinate directions in which they show. Hence any of these values can be used for the update of the mass properties in the frame of this research. As mentioned in chapter 5.4 the mass values are calculated for each element and then distributed equally to their grid points. Figure E.1 illustrates the global FEM grid points inside the slices along the wing which are set parallel to the streamlines.

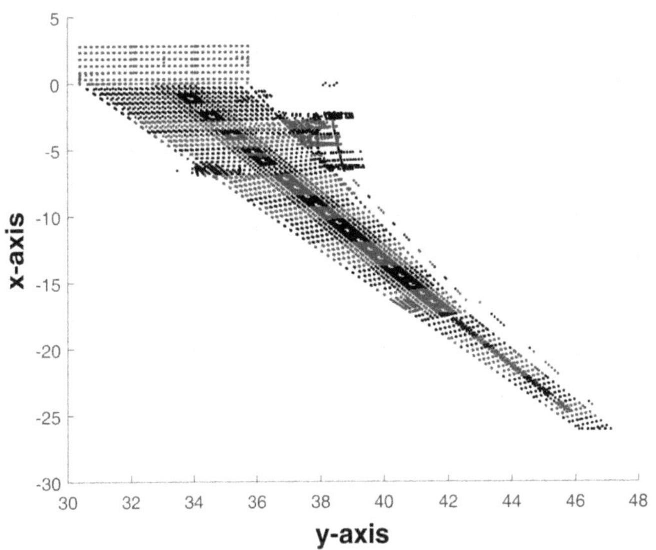

Figure E.1: Wing mass estimation approach - global FEM grid points and slices

F Influence of the load alleviation function on the calculated flight loads

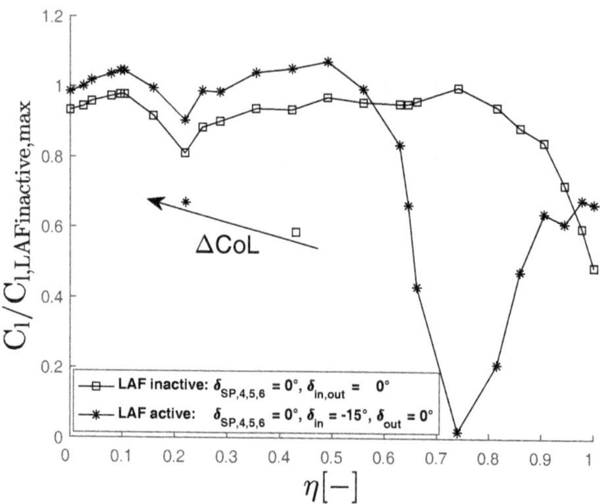

Figure F.1: Effect of inner and outer aileron upwards deflection on the normalized lift distribution of a steady flight case: $n_z = 1g$

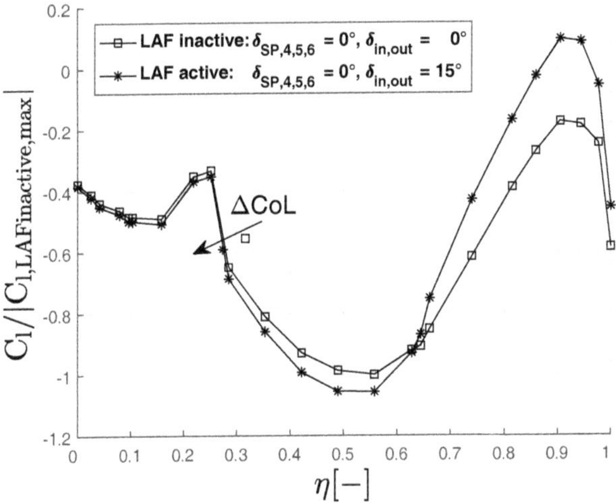

Figure F.2: Effect of inner and outer aileron downwards deflection on the normalized lift distribution of a push-down manoeuvre: $n_z = -1g$

G Structural baseline generation

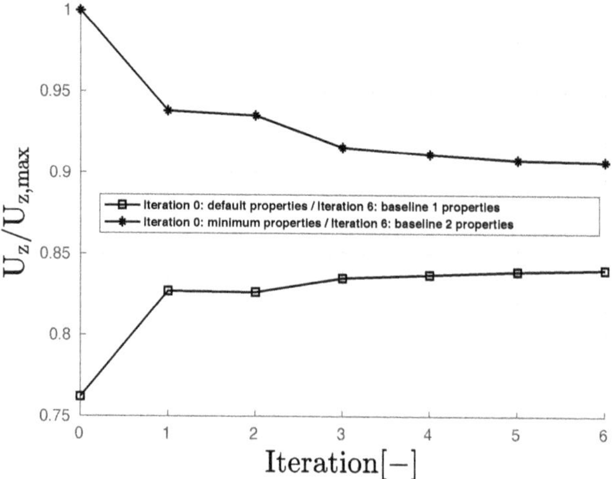

Figure G.1: Normalized wing tip vertical displacement U_z shown for six iteration loops starting with two different initial property sets and ending with two generated baseline models

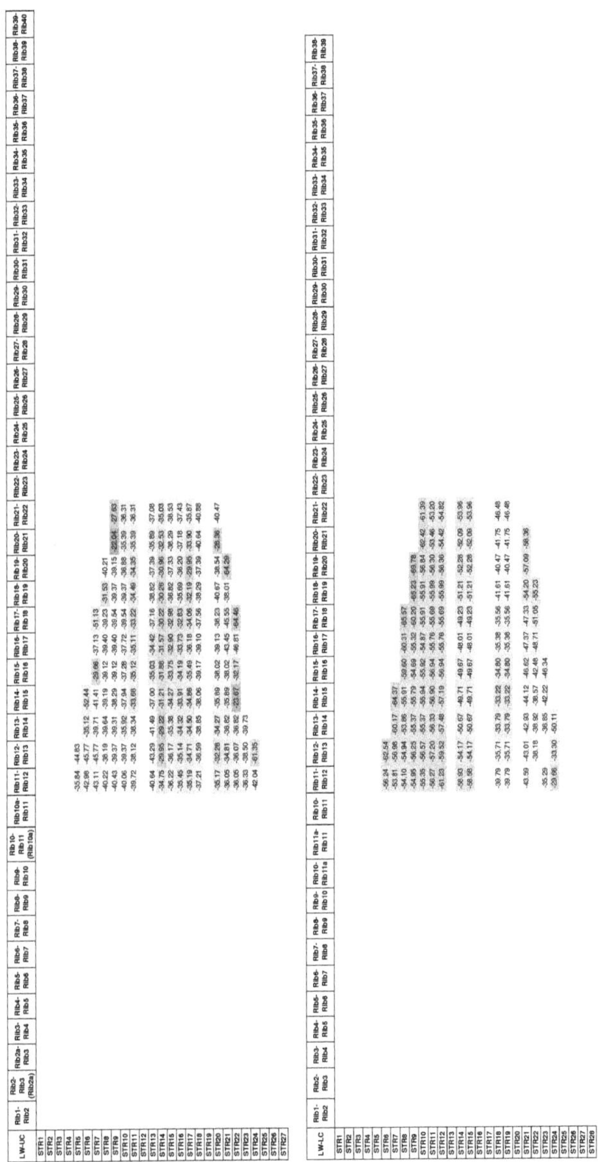

LW-UC	Rib1-Rib2	Rib2-Rib3 (Rib2a)	Rib2a-Rib3	Rib3-Rib4	Rib4-Rib5	Rib5-Rib6	Rib6-Rib7	Rib7-Rib8	Rib8-Rib9	Rib9-Rib10	Rib10-Rib11 (Rib10a)	Rib10a-Rib11	Rib11-Rib12	Rib12-Rib13	Rib13-Rib14	Rib14-Rib15	Rib15-Rib16	Rib16-Rib17	Rib17-Rib18	Rib18-Rib19	Rib19-Rib20	Rib20-Rib21	Rib21-Rib22	Rib22-Rib23	Rib23-Rib24	Rib24-Rib25	Rib25-Rib26	Rib26-Rib27	Rib27-Rib28	Rib28-Rib29	Rib29-Rib30	Rib30-Rib31	Rib31-Rib32	Rib32-Rib33	Rib33-Rib34	Rib34-Rib35	Rib35-Rib36	Rib36-Rib37	Rib37-Rib38	Rib38-Rib39	Rib39-Rib40
STR1																																									
STR2																																									
STR3																																									
STR4																																									
STR5													-35.84	-44.83																											
STR6													-42.98	-45.77	-35.12	-52.44																									
STR7													-43.11	-45.77	-39.71	-41.41	-29.66	-37.13	-51.13																						
STR8													-40.22	-38.19	-39.64	-39.19	-39.12	-39.40	-39.23	-31.53	-40.21																				
STR9													-40.43	-39.37	-39.31	-38.29	-39.12	-39.40	-39.54	-39.37	-39.15	-22.04	-27.63																		
STR10													-40.06	-39.37	-35.92	-37.94	-37.28	-37.72	-39.54	-39.37	-36.88	-35.39	-36.31																		
STR11													-39.72	-38.12	-38.34	-33.68	-35.12	-35.11	-33.22	-34.49	-34.35	-35.39	-36.31																		
STR12																																									
STR13													-40.64	-43.29	-41.49	-37.00	-35.03	-34.42	-37.16	-38.82	-37.39	-35.89	-37.08																		
STR14													-34.75	-29.95	-29.22	-31.21	-31.88	-31.57	-30.22	-30.26	-30.96	-32.53	-35.03																		
STR15													-36.22	-36.17	-35.38	-34.27	-33.75	-32.90	-32.98	-36.82	-37.33	-38.29	-38.53																		
STR16													-35.45	-35.14	-34.32	-33.91	-34.19	-33.73	-32.83	-35.69	-36.20	-37.18	-37.43																		
STR17													-35.19	-34.71	-34.50	-34.86	-35.49	-36.18	-34.06	-32.19	-29.95	-33.90	-35.87																		
STR18													-37.21	-36.59	-38.85	-38.06	-39.17	-39.10	-37.56	-38.29	-37.39	-40.64	-40.88																		
STR19																																									
STR20													-35.17	-32.28	-34.27	-35.89	-38.02	-39.13	-38.23	-40.67	-38.54	-28.36	-40.47																		
STR21													-36.05	-34.81	-36.82	-35.89	-38.02	-43.45	-45.55	-38.01	-64.29																				
STR22													-36.05	-36.07	-36.82	-23.67	-32.17	-46.81	-64.46																						
STR23													-36.33	-38.50	-39.73																										
STR24													-42.04	-61.35																											
STR25																																									
STR26																																									
STR27																																									

LW-LC	Rib1-Rib2	Rib2-Rib3	Rib3-Rib4	Rib4-Rib5	Rib5-Rib6	Rib6-Rib7	Rib7-Rib8	Rib8-Rib9	Rib9-Rib10	Rib10-Rib11a	Rib11a-Rib11	Rib10-Rib11	Rib11-Rib12	Rib12-Rib13	Rib13-Rib14	Rib14-Rib15	Rib15-Rib16	Rib16-Rib17	Rib17-Rib18	Rib18-Rib19	Rib19-Rib20	Rib20-Rib21	Rib21-Rib22	Rib22-Rib23	Rib23-Rib24	Rib24-Rib25	Rib25-Rib26	Rib26-Rib27	Rib27-Rib28	Rib28-Rib29	Rib29-Rib30	Rib30-Rib31	Rib31-Rib32	Rib32-Rib33	Rib33-Rib34	Rib34-Rib35	Rib35-Rib36	Rib36-Rib37	Rib37-Rib38	Rib38-Rib39
STR1																																								
STR2																																								
STR3																																								
STR4																																								
STR5																																								
STR6													-56.24	-62.54																										
STR7													-53.81	-56.98	-60.17	-64.37																								
STR8													-54.10	-54.94	-53.86	-55.91	-59.60	-60.31	-65.57																					
STR9													-54.95	-56.25	-55.37	-55.79	-54.69	-55.32	-60.20	-65.23	-69.78																			
STR10													-55.35	-56.57	-55.37	-55.94	-55.92	-54.87	-55.91	-55.91	-56.84	-62.42	-61.39																	
STR11													-56.27	-57.20	-56.33	-56.90	-56.94	-55.76	-55.69	-55.99	-56.30	-53.46	-53.20																	
STR12													-61.23	-59.52	-57.48	-57.19	-56.94	-55.76	-55.69	-55.99	-56.36	-54.42	-54.82																	
STR13																																								
STR14													-58.93	-54.17	-50.67	-49.71	-49.67	-48.01	-49.23	-51.21	-52.28	-52.09	-53.96																	
STR15													-58.58	-54.17	-50.67	-49.71	-49.67	-48.01	-49.23	-51.21	-52.28	-52.09	-53.96																	
STR16																																								
STR17																																								
STR18													-39.79	-35.71	-33.79	-33.22	-34.80	-35.38	-35.56	-41.61	-40.47	-41.75	-46.48																	
STR19													-39.79	-35.71	-33.79	-33.22	-34.80	-35.38	-35.56	-41.61	-40.47	-41.75	-46.48																	
STR20																																								
STR21													-43.59	-43.01	-42.93	-44.12	-46.62	-47.37	-47.33	-54.20	-57.09	-58.36																		
STR22														-38.18	-38.92	-38.57	-42.48	-48.71	-51.05	-55.23																				
STR23													-35.29		-36.85	-42.22	-46.34																							
STR24													-29.66	-33.30	-50.11																									
STR25																																								
STR26																																								
STR27																																								
STR28																																								

Figure G.2: Mapping of skin thickness percentage changes shown for wing baseline generation from default to baseline property values

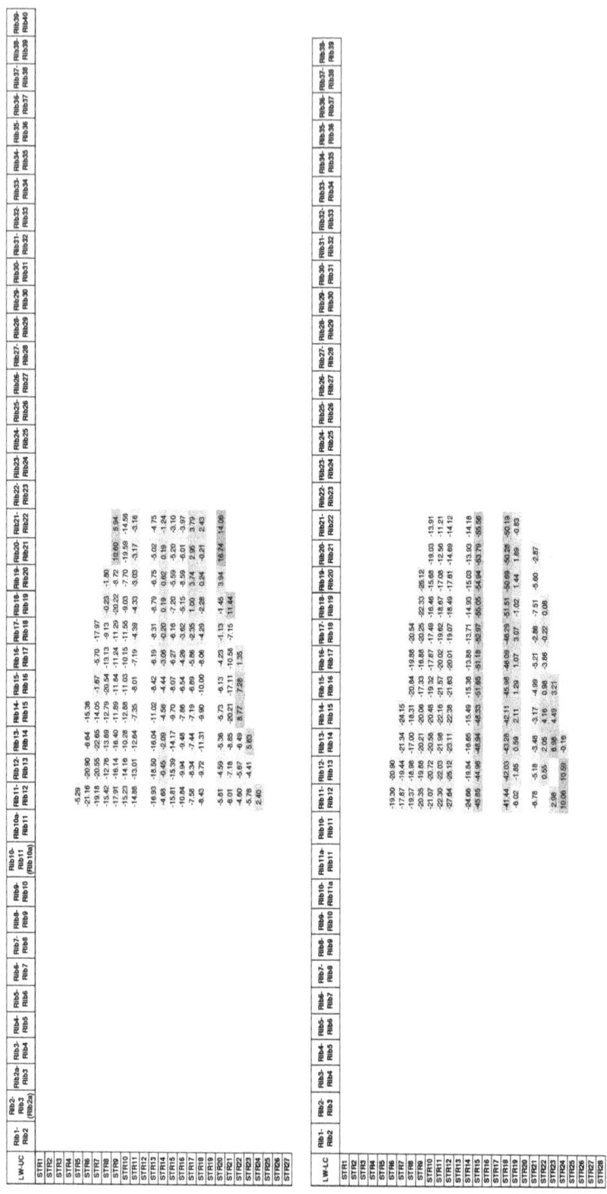

LW-UC	Rib1-Rib2	Rib2-Rib3 (Rib2a)	Rib2a-Rib3	Rib3-Rib4	Rib4-Rib5	Rib5-Rib6	Rib6-Rib7	Rib7-Rib8	Rib8-Rib9	Rib9-Rib10	Rib10-Rib11 (Rib10a)	Rib10a-Rib11
STR1												
STR2												
STR3												
STR4												
STR5												
STR6												
STR7												
STR8												
STR9												
STR10												
STR11												
STR12												
STR13												
STR14												
STR15												
STR16												
STR17												
STR18												
STR19												
STR20												
STR21												
STR22												
STR23												
STR24												
STR25												
STR26												
STR27												

LW-UC	Rib11-Rib12	Rib12-Rib13	Rib13-Rib14	Rib14-Rib15	Rib15-Rib16	Rib16-Rib17	Rib17-Rib18	Rib18-Rib19	Rib19-Rib20	Rib20-Rib21	Rib21-Rib22
STR1											
STR2											
STR3											
STR4											
STR5	-5.29										
STR6	-21.16	-20.90	-6.64	-15.38							
STR7	-19.18	-20.55	-22.65	-14.05	-1.67	-5.70	-17.97				
STR8	-15.42	-12.76	-13.69	-12.79	-20.54	-13.13	-9.13	-0.23	-1.80		
STR9	-17.91	-16.14	-16.40	-11.89	-11.84	-11.24	-11.29	-20.22	-8.72	10.60	5.94
STR10	-15.23	-14.16	-10.28	-12.88	-11.03	-10.15	-11.55	-9.03	-7.70	-19.59	-14.58
STR11	-14.88	-13.01	-12.64	-7.35	-8.01	-7.19	-4.39	-4.33	-3.03	-3.17	-3.16
STR12											
STR13	-16.93	-18.50	-16.04	-11.02	-8.42	-6.19	-8.31	-8.79	-6.75	-5.02	-4.75
STR14	-4.68	-0.45	-2.09	-4.56	-4.44	-3.06	-0.20	0.19	0.62	0.19	-1.24
STR15	-15.81	-15.39	-14.17	-9.70	-8.07	-6.27	-6.16	-7.20	-5.59	-5.20	-3.10
STR16	-10.84	-9.94	-9.48	-7.86	-6.54	-4.26	-3.62	-5.15	-8.59	-6.01	-3.97
STR17	-7.58	-8.34	-7.44	-7.19	-6.89	-5.86	-2.35	1.00	3.74	2.95	3.79
STR18	-8.43	-9.72	-11.31	-9.90	-10.00	-8.06	-4.29	-2.28	0.24	-0.21	2.43
STR19											
STR20	-5.81	-4.59	-5.36	-5.73	-6.13	-4.23	-1.13	-1.45	3.94	16.74	14.06
STR21	-6.01	-7.18	-8.85	-20.21	-17.11	-10.58	-7.15	11.44			
STR22	-4.60	-5.67	-6.49	8.77	7.28	1.35					
STR23	-5.76	-4.41	5.63								
STR24	2.40										
STR25											
STR26											
STR27											

LW-UC	Rib22-Rib23	Rib23-Rib24	Rib24-Rib25	Rib25-Rib26	Rib26-Rib27	Rib27-Rib28	Rib28-Rib29	Rib29-Rib30	Rib30-Rib31	Rib31-Rib32	Rib32-Rib33	Rib33-Rib34	Rib34-Rib35	Rib35-Rib36	Rib36-Rib37	Rib37-Rib38	Rib38-Rib39	Rib39-Rib40
STR1																		
STR2																		
STR3																		
STR4																		
STR5																		
STR6																		
STR7																		
STR8																		
STR9																		
STR10																		
STR11																		
STR12																		
STR13																		
STR14																		
STR15																		
STR16																		
STR17																		
STR18																		
STR19																		
STR20																		
STR21																		
STR22																		
STR23																		
STR24																		
STR25																		
STR26																		
STR27																		

LW-LC	Rib1-Rib2	Rib2-Rib3	Rib3-Rib4	Rib4-Rib5	Rib5-Rib6	Rib6-Rib7	Rib7-Rib8	Rib8-Rib9	Rib9-Rib10	Rib10-Rib11a	Rib11a-Rib11	Rib10-Rib11
STR1												
STR2												
STR3												
STR4												
STR5												
STR6												
STR7												
STR8												
STR9												
STR10												
STR11												
STR12												
STR13												
STR14												
STR15												
STR16												
STR17												
STR18												
STR19												
STR20												
STR21												
STR22												
STR23												
STR24												
STR25												
STR26												
STR27												
STR28												

LW-LC	Rib11-Rib12	Rib12-Rib13	Rib13-Rib14	Rib14-Rib15	Rib15-Rib16	Rib16-Rib17	Rib17-Rib18	Rib18-Rib19	Rib19-Rib20	Rib20-Rib21	Rib21-Rib22
STR1											
STR2											
STR3											
STR4											
STR5											
STR6	-19.30	-20.90									
STR7	-17.87	-19.44	-21.34	-24.15							
STR8	-19.37	-18.98	-17.00	-18.31	-20.84	-19.88	-20.54				
STR9	-20.35	-19.88	-20.21	-20.06	-17.33	-16.88	-20.25	-22.33	-25.12		
STR10	-21.07	-20.72	-20.58	-20.48	-19.32	-17.87	-17.49	-16.46	-15.68	-19.03	-13.91
STR11	-22.30	-22.03	-21.98	-22.16	-21.57	-20.02	-19.62	-18.67	-17.08	-12.56	-11.21
STR12	-27.64	-25.12	-23.11	-22.38	-21.63	-20.01	-19.07	-18.49	-17.61	-14.69	-14.12
STR13											
STR14	-24.66	-19.84	-16.65	-15.49	-15.36	-13.88	-13.71	-14.93	-15.03	-13.93	-14.18
STR15	-45.65	-44.96	-48.94	-48.33	-51.65	-51.18	-52.97	-55.05	-54.94	-53.79	-55.56
STR16											
STR17											
STR18	-41.44	-42.03	-43.28	-42.11	-45.98	-46.09	-46.29	-51.51	-50.69	-50.28	-50.19
STR19	-6.02	-1.85	0.59	2.11	1.29	1.07	3.07	-1.02	1.44	1.89	-0.83
STR20											
STR21	-6.78	-5.18	-3.48	-3.17	-4.99	-5.21	-2.86	-7.51	-5.60	-2.87	
STR22		0.55	2.05	4.16	0.98	-3.86	-0.22	0.08			
STR23	2.98		6.98	4.49	3.21						
STR24	10.06	10.59	-0.16								
STR25											
STR26											
STR27											
STR28											

LW-LC	Rib22-Rib23	Rib23-Rib24	Rib24-Rib25	Rib25-Rib26	Rib26-Rib27	Rib27-Rib28	Rib28-Rib29	Rib29-Rib30	Rib30-Rib31	Rib31-Rib32	Rib32-Rib33	Rib33-Rib34	Rib34-Rib35	Rib35-Rib36	Rib36-Rib37	Rib37-Rib38	Rib38-Rib39
STR1																	
STR2																	
STR3																	
STR4																	
STR5																	
STR6																	
STR7																	
STR8																	
STR9																	
STR10																	
STR11																	
STR12																	
STR13																	
STR14																	
STR15																	
STR16																	
STR17																	
STR18																	
STR19																	
STR20																	
STR21																	
STR22																	
STR23																	
STR24																	
STR25																	
STR26																	
STR27																	
STR28																	

Figure G.3: Mapping of stringer cross section area percentage changes shown for wing baseline generation from default to baseline property values

H Influence of load alleviation function on structural components

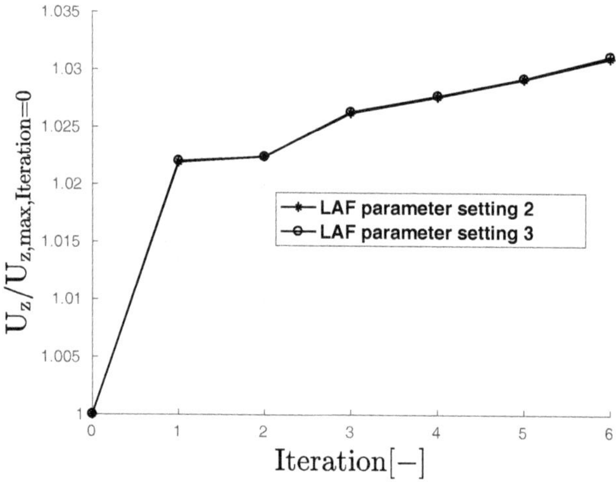

Figure H.1: Change of normalized wing tip vertical displacement U_z shown for two different LAF parameter settings

LW-UC	Rib1-Rib2	Rib2-Rib3 (Rib2a)	Rib2a-Rib3	Rib3-Rib4	Rib4-Rib5	Rib5-Rib6	Rib6-Rib7	Rib7-Rib8	Rib8-Rib9	Rib9-Rib10	Rib10-Rib11 (Rib10a)	Rib10a-Rib11	Rib11-Rib12	Rib12-Rib13	Rib13-Rib14	Rib14-Rib15	Rib15-Rib16	Rib16-Rib17	Rib17-Rib18	Rib18-Rib19	Rib19-Rib20	Rib20-Rib21	Rib21-Rib22	Rib22-Rib23	Rib23-Rib24	Rib24-Rib25	Rib25-Rib26	Rib26-Rib27	Rib27-Rib28	Rib28-Rib29	Rib29-Rib30	Rib30-Rib31	Rib31-Rib32	Rib32-Rib33	Rib33-Rib34	Rib34-Rib35	Rib35-Rib36	Rib36-Rib37	Rib37-Rib38	Rib38-Rib39	Rib39-Rib40
STR1																																								
STR2																																								
STR3																																								
STR4																																								
STR5													-15.72	-17.79																										
STR6													-13.92	-14.80	-16.96	-18.79																								
STR7													-13.92	-14.80	-15.55	-16.38	-16.36	-15.05	-15.52																					
STR8													-13.89	-14.89	-15.55	-15.44	-13.30	-12.45	-12.70	-12.12	-11.60																			
STR9													-13.89	-15.29	-15.16	-14.54	-13.30	-12.45	-11.55	-10.37	-9.37	-7.97	-6.26																	
STR10													-13.92	-15.29	-14.84	-14.26	-13.32	-12.45	-11.55	-10.37	-9.37	-6.91	-5.58																	
STR11													-14.46	-14.87	-14.01	-13.20	-12.46	-11.44	-10.53	-9.10	-8.08	-6.91	-5.63																	
STR12																																								
STR13													-15.30	-15.91	-15.59	-14.52	-13.42	-12.38	-11.70	-10.72	-9.60	-8.06	-6.56																	
STR14													-13.83	-13.07	-12.20	-11.75	-10.96	-10.08	-8.87	-7.91	-6.73	-5.67	-4.77																	
STR15													-13.50	-13.43	-12.43	-11.42	-10.36	-9.52	-8.58	-8.13	-7.19	-5.78	-4.67																	
STR16													-13.50	-13.43	-12.43	-11.18	-10.15	-9.17	-7.93	-8.13	-7.19	-5.78	-4.67																	
STR17													-13.59	-12.92	-12.06	-11.35	-10.26	-9.19	-7.97	-6.72	-5.72	-4.34	-3.48																	
STR18													-14.26	-13.90	-12.89	-11.71	-10.41	-9.48	-8.61	-7.27	-6.78	-5.14	-4.93																	
STR19																																								
STR20													-13.52	-13.42	-12.16	-11.00	-9.57	-8.57	-7.22	-6.38	-4.59	-4.72	-4.08																	
STR21													-13.52	-13.73	-12.71	-10.91	-9.57	-7.64	-7.39	-5.60	0.00																			
STR22													-13.52	-13.64	-12.71	-10.59	-9.56	-7.77	0.85																					
STR23													-13.31	-13.97	-12.95																									
STR24													-15.47	-15.43																										
STR25																																								
STR26																																								
STR27																																								

LW-LC	Rib1-Rib2	Rib2-Rib3	Rib3-Rib4	Rib4-Rib5	Rib5-Rib6	Rib6-Rib7	Rib7-Rib8	Rib8-Rib9	Rib9-Rib10	Rib10-Rib11a	Rib11a-Rib11	Rib10-Rib11	Rib11-Rib12	Rib12-Rib13	Rib13-Rib14	Rib14-Rib15	Rib15-Rib16	Rib16-Rib17	Rib17-Rib18	Rib18-Rib19	Rib19-Rib20	Rib20-Rib21	Rib21-Rib22	Rib22-Rib23	Rib23-Rib24	Rib24-Rib25	Rib25-Rib26	Rib26-Rib27	Rib27-Rib28	Rib28-Rib29	Rib29-Rib30	Rib30-Rib31	Rib31-Rib32	Rib32-Rib33	Rib33-Rib34	Rib34-Rib35	Rib35-Rib36	Rib36-Rib37	Rib37-Rib38	Rib38-Rib39
STR1																																							
STR2																																							
STR3																																							
STR4																																							
STR5																																							
STR6													-24.48	-24.04																									
STR7													-30.69	-27.22	-34.92	-33.63																							
STR8													-32.55	-30.46	-28.58	-28.83	-31.95	-35.59	-35.87																				
STR9													-33.72	-34.47	-33.20	-31.55	-28.38	-28.39	-31.56	-36.78	-28.93																		
STR10													-33.77	-34.47	-33.20	-33.26	-31.92	-27.52	-23.95	-20.22	-19.08	-16.64	-23.45																
STR11													-34.88	-37.76	-38.26	-37.65	-35.11	-32.70	-29.84	-26.62	-19.54	-14.35	-11.50																
STR12													-40.18	-42.58	-40.70	-38.12	-35.19	-32.70	-29.84	-26.62	-25.40	-22.17	-17.15																
STR13																																							
STR14													-41.40	-40.42	-41.08	-40.38	-39.18	-37.42	-39.37	-39.45	-40.00	-35.80	-40.79																
STR15													-18.35	-22.66	-14.15	-7.66	-0.80	2.75	14.05	20.05	24.87	21.18	35.10																
STR16																																							
STR17																																							
STR18													8.01	10.95	10.50	13.11	12.13	13.82	17.41	24.53	19.89	15.07	13.26																
STR19													-50.28	-52.18	-48.50	-44.98	-39.79	-36.64	-36.17	-37.31	-30.99	-26.25	-23.15																
STR20																																							
STR21													-40.23	-43.19	-45.59	-47.91	-44.78	-40.93	-37.70	-37.99	-32.54	-18.06																	
STR22														-36.60	-38.24	-38.77	-43.14	-42.38	-41.97	-32.21																			
STR23													-33.69		-32.01	-30.54	-27.87																						
STR24													-28.25	-29.58	-29.27																								
STR25																																							
STR26																																							
STR27																																							
STR28																																							

Figure H.2: Mapping of skin thickness percentage changes shown for the wing baseline model and the applied LAF parameter setting 2

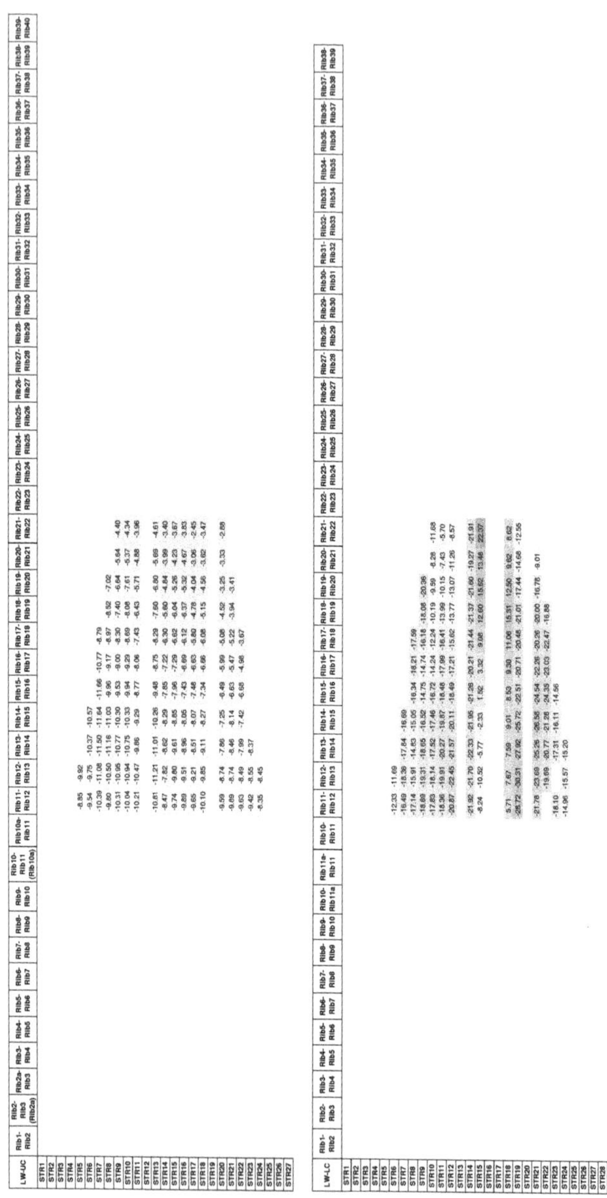

LW-UC	Rib1-Rib2	Rib2-Rib3 (Rib2a)	Rib2a-Rib3	Rib3-Rib4	Rib4-Rib5	Rib5-Rib6	Rib6-Rib7	Rib7-Rib8	Rib8-Rib9	Rib9-Rib10	Rib10-Rib11 (Rib10a)	Rib10a-Rib11	Rib11-Rib12	Rib12-Rib13	Rib13-Rib14	Rib14-Rib15	Rib15-Rib16	Rib16-Rib17	Rib17-Rib18	Rib18-Rib19	Rib19-Rib20	Rib20-Rib21	Rib21-Rib22	Rib22-Rib23	Rib23-Rib24	Rib24-Rib25	Rib25-Rib26	Rib26-Rib27	Rib27-Rib28	Rib28-Rib29	Rib29-Rib30	Rib30-Rib31	Rib31-Rib32	Rib32-Rib33	Rib33-Rib34	Rib34-Rib35	Rib35-Rib36	Rib36-Rib37	Rib37-Rib38	Rib38-Rib39	Rib39-Rib40
STR1																																									
STR2																																									
STR3																																									
STR4																																									
STR5													-8.85	-9.92																											
STR6													-9.54	-9.75	-10.37	-10.57																									
STR7													-10.39	-11.08	-11.50	-11.64	-11.66	-10.77	-8.79																						
STR8													-9.80	-10.50	-11.16	-11.03	-9.96	-9.17	-8.97	-8.52	-7.02																				
STR9													-10.31	-10.95	-10.77	-10.30	-9.53	-9.00	-8.30	-7.40	-6.64	-5.64	-4.40																		
STR10													-10.04	-10.94	-10.75	-10.33	-9.94	-9.29	-8.69	-8.08	-7.61	-5.37	-4.34																		
STR11													-10.21	-10.47	-9.86	-9.29	-8.77	-8.06	-7.43	-6.43	-5.71	-4.88	-3.96																		
STR12																																									
STR13													-10.81	-11.21	-11.01	-10.26	-9.48	-8.75	-8.29	-7.60	-6.80	-5.69	-4.61																		
STR14													-8.47	-7.82	-8.62	-8.29	-7.85	-7.22	-6.30	-5.60	-4.84	-3.99	-3.40																		
STR15													-9.74	-9.80	-9.61	-8.85	-7.96	-7.29	-6.62	-6.04	-5.26	-4.23	-3.67																		
STR16													-9.89	-9.51	-8.96	-8.05	-7.43	-6.69	-6.12	-6.37	-5.32	-4.67	-3.83																		
STR17													-9.65	-9.21	-8.51	-8.07	-7.48	-6.63	-5.80	-4.78	-4.04	-3.06	-2.45																		
STR18													-10.10	-9.85	-9.11	-8.27	-7.34	-6.66	-6.08	-5.15	-4.56	-3.62	-3.47																		
STR19																																									
STR20													-9.59	-8.74	-7.86	-7.25	-6.49	-5.99	-5.08	-4.52	-3.25	-3.33	-2.88																		
STR21													-9.89	-8.74	-8.46	-8.14	-6.63	-5.47	-5.22	-3.94	-3.41																				
STR22													-9.63	-8.49	-7.99	-7.42	-6.68	-4.98	-3.67																						
STR23													-9.42	-8.55	-8.37																										
STR24													-8.35	-8.45																											
STR25																																									
STR26																																									
STR27																																									

LW-LC	Rib1-Rib2	Rib2-Rib3	Rib3-Rib4	Rib4-Rib5	Rib5-Rib6	Rib6-Rib7	Rib7-Rib8	Rib8-Rib9	Rib9-Rib10	Rib10-Rib11a	Rib11a-Rib11	Rib10-Rib11	Rib11-Rib12	Rib12-Rib13	Rib13-Rib14	Rib14-Rib15	Rib15-Rib16	Rib16-Rib17	Rib17-Rib18	Rib18-Rib19	Rib19-Rib20	Rib20-Rib21	Rib21-Rib22	Rib22-Rib23	Rib23-Rib24	Rib24-Rib25	Rib25-Rib26	Rib26-Rib27	Rib27-Rib28	Rib28-Rib29	Rib29-Rib30	Rib30-Rib31	Rib31-Rib32	Rib32-Rib33	Rib33-Rib34	Rib34-Rib35	Rib35-Rib36	Rib36-Rib37	Rib37-Rib38	Rib38-Rib39
STR1																																							
STR2																																							
STR3																																							
STR4																																							
STR5																																							
STR6													-12.33	-11.69																									
STR7													-16.49	-18.36	-17.84	-16.69																							
STR8													-17.14	-15.91	-14.83	-15.05	-16.34	-16.21	-17.59																				
STR9													-18.69	-19.31	-18.65	-16.52	-14.75	-14.74	-16.18	-18.08	-20.36																		
STR10													-17.83	-18.14	-17.52	-17.46	-16.72	-14.24	-12.24	-10.19	-9.59	-8.28	-11.68																
STR11													-18.36	-19.91	-20.27	-19.87	-18.48	-17.99	-16.41	-13.99	-10.15	-7.43	-5.70																
STR12													-20.87	-22.45	-21.57	-20.11	-18.49	-17.21	-15.62	-13.77	-13.07	-11.26	-8.57																
STR13																																							
STR14													-21.92	-21.70	-22.33	-21.95	-21.26	-20.21	-21.44	-21.37	-21.60	-19.27	-21.91																
STR15													-8.24	-10.52	-5.77	-2.33	1.52	3.32	9.08	12.60	15.62	13.48	22.37																
STR16																																							
STR17																																							
STR18													5.71	7.67	7.59	9.01	8.53	9.30	11.06	15.31	12.50	9.62	8.62																
STR19													-28.72	-30.31	-27.92	-25.72	-22.51	-20.71	-20.48	-21.01	-17.44	-14.68	-12.55																
STR20																																							
STR21													-21.78	-23.69	-25.26	-26.58	-24.54	-22.26	-20.26	-20.00	-16.78	-9.01																	
STR22														-19.69	-20.77	-21.28	-24.35	-23.03	-22.47	-16.88																			
STR23													-18.10		-17.31	-16.11	-14.56																						
STR24													-14.96	-15.57	-15.20																								
STR25																																							
STR26																																							
STR27																																							
STR28																																							

Figure H.3: Mapping of stringer cross section area percentage changes shown for the wing baseline model and the applied LAF parameter setting 2

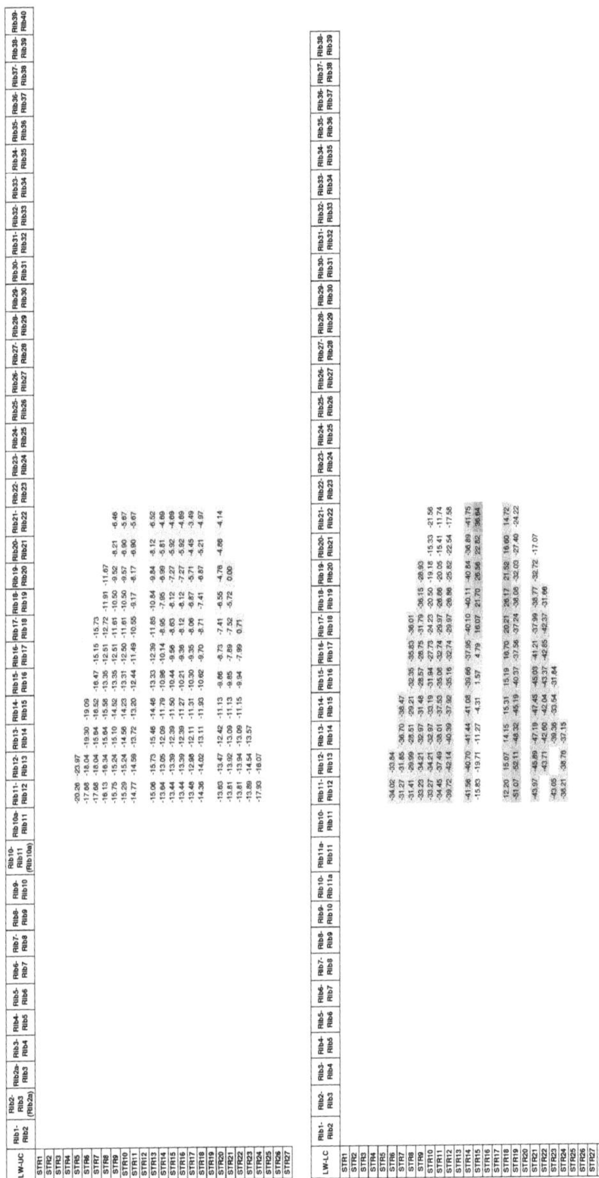

LW-UC	Rib1-Rib2	Rib2-Rib3 (Rib2a)	Rib2a-Rib3	Rib3-Rib4	Rib4-Rib5	Rib5-Rib6	Rib6-Rib7	Rib7-Rib8	Rib8-Rib9	Rib9-Rib10	Rib10-Rib11 (Rib10a)	Rib10a-Rib11	Rib11-Rib12	Rib12-Rib13	Rib13-Rib14	Rib14-Rib15	Rib15-Rib16	Rib16-Rib17	Rib17-Rib18	Rib18-Rib19	Rib19-Rib20	Rib20-Rib21	Rib21-Rib22	Rib22-Rib23	Rib23-Rib24	Rib24-Rib25	Rib25-Rib26	Rib26-Rib27	Rib27-Rib28	Rib28-Rib29	Rib29-Rib30	Rib30-Rib31	Rib31-Rib32	Rib32-Rib33	Rib33-Rib34	Rib34-Rib35	Rib35-Rib36	Rib36-Rib37	Rib37-Rib38	Rib38-Rib39	Rib39-Rib40
STR1																																									
STR2																																									
STR3																																									
STR4																																									
STR5													-20.26	-23.97																											
STR6													-17.68	-18.04	-19.30	-19.09																									
STR7													-17.68	-18.04	-15.64	-16.52	-16.47	-15.15	-15.73																						
STR8													-16.13	-16.34	-15.64	-15.58	-13.35	-12.51	-12.72	-11.91	-11.67																				
STR9													-15.75	-15.24	-15.10	-14.52	-13.35	-12.51	-11.61	-10.50	-9.52	-8.21	-6.46																		
STR10													-15.29	-15.24	-14.56	-14.23	-13.31	-12.50	-11.61	-10.50	-9.57	-6.90	-5.67																		
STR11													-14.77	-14.59	-13.72	-13.20	-12.44	-11.49	-10.55	-9.17	-8.17	-6.90	-5.67																		
STR12																																									
STR13													-15.06	-15.73	-15.46	-14.46	-13.33	-12.39	-11.85	-10.84	-9.84	-8.12	-6.52																		
STR14													-13.64	-13.05	-12.09	-11.79	-10.96	-10.14	-8.95	-7.95	-6.99	-5.81	-4.69																		
STR15													-13.44	-13.39	-12.39	-11.50	-10.44	-9.56	-8.63	-8.12	-7.27	-5.92	-4.69																		
STR16													-13.44	-13.39	-12.39	-11.27	-10.21	-9.36	-8.12	-8.12	-7.27	-5.92	-4.69																		
STR17													-13.48	-12.98	-12.11	-11.31	-10.30	-9.35	-8.06	-6.87	-5.71	-4.45	-3.49																		
STR18													-14.36	-14.02	-13.11	-11.93	-10.62	-9.70	-8.71	-7.41	-6.87	-5.21	-4.97																		
STR19																																									
STR20													-13.63	-13.47	-12.42	-11.13	-9.86	-8.73	-7.41	-6.55	-4.76	-4.86	-4.14																		
STR21													-13.81	-13.92	-13.09	-11.13	-9.85	-7.89	-7.52	-5.72	0.00																				
STR22													-13.81	-13.94	-13.09	-11.15	-9.94	-7.99	0.71																						
STR23													-13.89	-14.54	-13.57																										
STR24													-17.93	-16.07																											
STR25																																									
STR26																																									
STR27																																									

LW-LC	Rib1-Rib2	Rib2-Rib3	Rib3-Rib4	Rib4-Rib5	Rib5-Rib6	Rib6-Rib7	Rib7-Rib8	Rib8-Rib9	Rib9-Rib10	Rib10-Rib11a	Rib11a-Rib11	Rib10-Rib11	Rib11-Rib12	Rib12-Rib13	Rib13-Rib14	Rib14-Rib15	Rib15-Rib16	Rib16-Rib17	Rib17-Rib18	Rib18-Rib19	Rib19-Rib20	Rib20-Rib21	Rib21-Rib22	Rib22-Rib23	Rib23-Rib24	Rib24-Rib25	Rib25-Rib26	Rib26-Rib27	Rib27-Rib28	Rib28-Rib29	Rib29-Rib30	Rib30-Rib31	Rib31-Rib32	Rib32-Rib33	Rib33-Rib34	Rib34-Rib35	Rib35-Rib36	Rib36-Rib37	Rib37-Rib38	Rib38-Rib39
STR1																																								
STR2																																								
STR3																																								
STR4																																								
STR5																																								
STR6													-34.02	-33.84																										
STR7													-31.27	-31.85	-36.70	-38.47																								
STR8													-31.41	-29.99	-28.51	-29.21	-32.35	-35.83	-36.01																					
STR9													-33.23	-34.21	-32.97	-31.48	-28.57	-28.75	-31.79	-36.15	-28.93																			
STR10													-33.27	-34.21	-32.97	-33.19	-31.94	-27.73	-24.23	-20.50	-19.18	-15.33	-21.56																	
STR11													-34.45	-37.49	-38.01	-37.53	-35.06	-32.74	-29.97	-26.86	-20.05	-15.41	-11.74																	
STR12													-39.72	-42.14	-40.39	-37.92	-35.16	-32.74	-29.97	-26.86	-25.82	-22.54	-17.58																	
STR13																																								
STR14													-41.56	-40.70	-41.44	-41.08	-39.66	-37.95	-40.10	-40.11	-40.84	-36.89	-41.75																	
STR15													-15.83	-19.71	-11.27	-4.31	1.57	4.79	16.07	21.70	26.56	22.82	36.64																	
STR16																																								
STR17																																								
STR18													12.20	15.07	14.15	15.31	15.19	16.70	20.21	26.17	21.52	16.60	14.72																	
STR19													-51.07	-52.11	-48.32	-45.19	-40.37	-37.56	-37.24	-38.06	-32.03	-27.40	-24.22																	
STR20																																								
STR21													-43.97	-45.89	-47.19	-47.45	-45.03	-41.21	-37.99	-38.77	-32.72	-17.07																		
STR22														-43.71	-42.60	-42.04	-43.37	-42.85	-42.37	-31.66																				
STR23													-43.05		-39.36	-33.54	-31.84																							
STR24													-38.21	-38.76	-37.15																									
STR25																																								
STR26																																								
STR27																																								
STR28																																								

Figure H.4: Mapping of skin thickness percentage changes shown for the wing baseline model and the applied LAF parameter setting 3

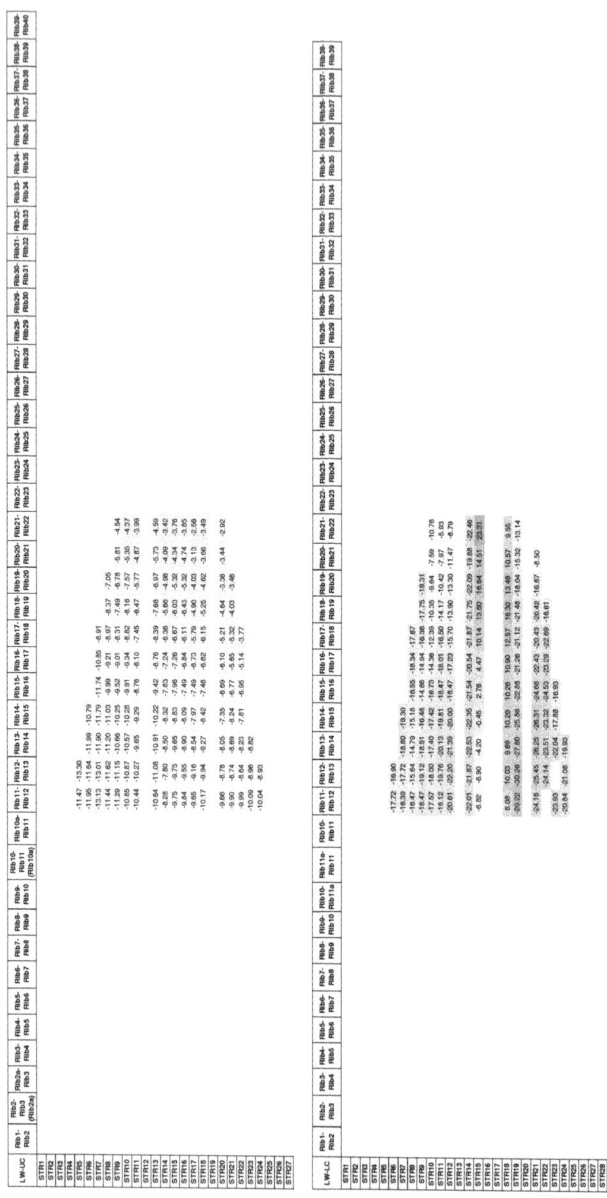

LW-UC	Rib1-Rib2	Rib2-Rib3 (Rib2a)	Rib2a-Rib3	Rib3-Rib4	Rib4-Rib5	Rib5-Rib6	Rib6-Rib7	Rib7-Rib8	Rib8-Rib9	Rib9-Rib10	Rib10-Rib11 (Rib10a)	Rib10a-Rib11	Rib11-Rib12	Rib12-Rib13	Rib13-Rib14	Rib14-Rib15	Rib15-Rib16	Rib16-Rib17	Rib17-Rib18	Rib18-Rib19	Rib19-Rib20	Rib20-Rib21	Rib21-Rib22	Rib22-Rib23	Rib23-Rib24	Rib24-Rib25	Rib25-Rib26	Rib26-Rib27	Rib27-Rib28	Rib28-Rib29	Rib29-Rib30	Rib30-Rib31	Rib31-Rib32	Rib32-Rib33	Rib33-Rib34	Rib34-Rib35	Rib35-Rib36	Rib36-Rib37	Rib37-Rib38	Rib38-Rib39	Rib39-Rib40
STR1																																									
STR2																																									
STR3																																									
STR4																																									
STR5													-11.47	-13.30																											
STR6													-11.95	-11.64	-11.99	-10.79																									
STR7													-13.13	-13.01	-11.90	-11.79	-11.74	-10.85	-8.91																						
STR8													-11.44	-11.62	-11.20	-11.03	-9.99	-9.21	-8.97	-8.37	-7.05																				
STR9													-11.29	-11.15	-10.66	-10.25	-9.52	-9.01	-8.31	-7.49	-6.78	-5.81	-4.54																		
STR10													-10.85	-10.87	-10.57	-10.28	-9.91	-9.34	-8.82	-8.16	-7.57	-5.35	-4.37																		
STR11													-10.44	-10.27	-9.65	-9.29	-8.76	-8.10	-7.45	-6.47	-5.77	-4.87	-3.99																		
STR12																																									
STR13													-10.64	-11.08	-10.91	-10.22	-9.42	-8.76	-8.39	-7.68	-6.97	-5.73	-4.59																		
STR14													-8.28	-7.80	-8.50	-8.32	-7.83	-7.24	-6.36	-5.86	-4.98	-4.09	-3.42																		
STR15													-9.75	-9.75	-9.65	-8.83	-7.96	-7.26	-6.67	-6.03	-5.32	-4.34	-3.76																		
STR16													-9.84	-9.55	-8.90	-8.09	-7.49	-6.84	-6.11	-6.43	-5.32	-4.74	-3.85																		
STR17													-9.65	-9.16	-8.54	-7.97	-7.49	-6.73	-5.79	-4.90	-4.03	-3.13	-2.56																		
STR18													-10.17	-9.94	-9.27	-8.42	-7.48	-6.82	-6.15	-5.25	-4.62	-3.66	-3.49																		
STR19																																									
STR20													-9.66	-8.78	-8.05	-7.35	-6.69	-6.10	-5.21	-4.64	-3.38	-3.44	-2.92																		
STR21													-9.90	-8.74	-8.69	-8.24	-6.77	-5.65	-5.32	-4.03	-3.46																				
STR22													-9.99	-8.64	-8.23	-7.81	-6.95	-5.14	-3.77																						
STR23													-10.09	-8.96	-8.82																										
STR24													-10.04	-8.93																											
STR25																																									
STR26																																									
STR27																																									

LW-LC	Rib1-Rib2	Rib2-Rib3	Rib3-Rib4	Rib4-Rib5	Rib5-Rib6	Rib6-Rib7	Rib7-Rib8	Rib8-Rib9	Rib9-Rib10	Rib10-Rib11a	Rib11a-Rib11	Rib10-Rib11	Rib11-Rib12	Rib12-Rib13	Rib13-Rib14	Rib14-Rib15	Rib15-Rib16	Rib16-Rib17	Rib17-Rib18	Rib18-Rib19	Rib19-Rib20	Rib20-Rib21	Rib21-Rib22	Rib22-Rib23	Rib23-Rib24	Rib24-Rib25	Rib25-Rib26	Rib26-Rib27	Rib27-Rib28	Rib28-Rib29	Rib29-Rib30	Rib30-Rib31	Rib31-Rib32	Rib32-Rib33	Rib33-Rib34	Rib34-Rib35	Rib35-Rib36	Rib36-Rib37	Rib37-Rib38	Rib38-Rib39
STR1																																								
STR2																																								
STR3																																								
STR4																																								
STR5																																								
STR6													-17.72	-16.90																										
STR7													-16.39	-17.72	-18.80	-19.30																								
STR8													-16.47	-15.64	-14.79	-15.18	-16.55	-18.34	-17.67																					
STR9													-18.47	-19.12	-18.51	-16.48	-14.86	-14.94	-16.36	-17.75	-19.31																			
STR10													-17.57	-18.00	-17.40	-17.42	-16.73	-14.36	-12.39	-10.35	-9.64	-7.59	-10.76																	
STR11													-18.12	-19.76	-20.13	-19.81	-18.47	-18.01	-16.50	-14.17	-10.42	-7.97	-5.93																	
STR12													-20.61	-22.20	-21.39	-20.00	-18.47	-17.23	-15.70	-13.90	-13.30	-11.47	-8.79																	
STR13																																								
STR14													-22.01	-21.87	-22.53	-22.35	-21.54	-20.54	-21.87	-21.75	-22.09	-19.88	-22.46																	
STR15													-6.82	-8.90	-4.20	-0.45	2.76	4.47	10.14	13.60	16.64	14.51	23.31																	
STR16																																								
STR17																																								
STR18													8.08	10.03	9.69	10.20	10.26	10.90	12.57	16.30	13.48	10.57	9.55																	
STR19													-29.22	-30.26	-27.80	-25.86	-22.88	-21.26	-21.12	-21.48	-18.04	-15.32	-13.14																	
STR20																																								
STR21													-24.16	-25.45	-26.25	-26.31	-24.66	-22.43	-20.43	-20.42	-16.87	-8.50																		
STR22														-24.14	-23.51	-23.32	-24.53	-23.29	-22.69	-16.61																				
STR23													-23.93		-22.04	-17.88	-16.93																							
STR24													-20.84	-21.08	-19.93																									
STR25																																								
STR26																																								
STR27																																								
STR28																																								

Figure H.5: Mapping of stringer cross section area percentage changes shown for the wing baseline model and the applied LAF parameter setting 3

www.ingramcontent.com/pod-product-compliance
Ingram Content Group UK Ltd.
Pitfield, Milton Keynes, MK11 3LW, UK
UKHW022001190726
13853UKWH00004B/1659

9 783736 975163